After Access

The Information Society Series
Laura DeNardis and Michael Zimmer, series editors

After Access

Inclusion, Development, and a More Mobile Internet

Jonathan Donner

WITHDRAWN

The MIT Press
Cambridge, Massachusetts
London, England

This book was set in Stone Sans and Stone Serif by Toppan Best-set Premedia Limited. Printed and bound in the United States of America.

Library of Congress Cataloging-in-Publication Data is available.

ISBN: 978-0-262-02992-6

10 9 8 7 6 5 4 3 2 1

To Gary and Theo, wishing you two could have met.

Contents

Acknowledgments

Time-stamped emails on my PC suggest I started thinking about writing this book in 2012. However, it reflects and contains work going back much further, spanning a longer engagement in the ICT4D and mobile communication research communities. It is an honor to be a member of each, and I am grateful for over a decade of discussions around the world.

Many colleagues at Microsoft Research, including Rosa Arriaga, Melissa Densmore, Mary Gray, Richard Harper, Indrani Medhi, Jacki O'Neill, Nimmi Rangaswamy, and Bill Thies, engaged with drafts and strengthened my arguments. My managers, Kentaro Toyama, Ed Cutrell, and P. Anandan, were remarkably supportive of my writing efforts. Likewise, my new colleagues at Caribou Digital, Savita Bailur, Marissa Drouillard, Tim Hayward, Chris Locke, Kishor Nagula, Bryan Pon, and Emrys Schoemaker, have been patient as I revised the document in response to the great feedback from the anonymous peer reviewers assigned by the MIT Press. Thanks also to the team at the MIT Press, Margy Avery, Susan Buckley, Julia Collins, Virginia Crossman, and Gita Manaktala, for all their assistance.

The book draws on several studies I conducted with coauthors: Richard Banks, A. J. Brush, Marshini Chetty, Shikoh Gitau, Beki Grinter, Andrew Maunder, Preeti Mudliar, Bill Thies, and Marion Walton, as well on ideas in my previous books with Richard Ling and Patricia Mechael. I am grateful to all of them, and hope I have done justice to our collaborative work.

Several presentations around the world let me explore initial ideas with students and colleagues. Scott Campbell, Nicola Dell, Marcela Escobari, Richard Heeks, James Katz, Monroe Price, Laura Stark, and Wayan Vota were great hosts. Special thanks to Katy Pearce, not only a host but also a particularly close reader of a full draft of the book. Her attention to details and the big picture improved this book tremendously.

Twitter and the blogosphere kept me abreast of the changing technology landscape. Google Scholar found literature. OneNote kept things tidy, Mendeley wrangled citations, Word put it all together, and a serendipitous Foursquare check-in helped kick the whole thing off. To balance all this typing and tapping, Katya Kinski, Jackie Wells, and Tracy Bertish kept me healthy and able to work.

Of course and as always, Caitlin, Calliope, and Theo supported in every way possible, listening to ideas turning over in my head when I should have been stopping to smell the fynbos, and forgiving my long absences secluded away at the keyboard. I am grateful for every day I have with them.

Finally, I am sorry not to be able to thank Gary Marsden in person for all he did to support this work. Gary was a great friend and the most generous and enthusiastic colleague a researcher could hope to meet. He welcomed me to South Africa, was a coauthor on the original "After Access" paper, and the inspiration for many of the ideas in this book. We miss him very much.

I   The Boom

1 Introduction

When describing mobile phone use around the world, many writers start with a graphic illustrating the rapid growth in *something* over the last ten or twenty years—users, subscriptions, coverage, devices, even revenues. I am not immune to this temptation. Richard Ling and I were only five pages into our 2009 book, *Mobile Communication,* before we added a graphic depicting the steep rise in mobile telephone subscriptions that overtook worldwide landline subscriptions around the turn of the millennium.[1] One could certainly offer a similar graphic today, depicting the shift from an Internet dominated by personal computers (PCs) and wired connections to one teeming with mobile devices connected by wireless signals.

Such shifts—from fixed telephony to mobile, and more recently from fixed Internet toward mobile Internet—make for great stage-setting initial paragraphs because they are palpable and exciting. We are all participants in the global mobile boom. Since that 2009 book, over a billion more people have become mobile technology users, not only in the prosperous Global North, but also in the cities and villages of the Global South, where access to telecommunications services and networks has traditionally been scarcer and more expensive. Mobile devices (and signals) are not yet everywhere, but thanks to the mobile boom, the vast majority of adults around the world will soon have access to a phone, and by extension, an Internet connection.

The boom has promoted a commensurate outpouring of enthusiasm from the technology, policy, and development communities. For example, noted economist Jeffrey Sachs has called the mobile phone "the single most transformative technology for development."[2] The World Bank's flagship publication on information and communication technologies for development (ICT4D) makes a similar claim: "Mobile communication has arguably had a bigger impact on humankind in a shorter period of time than any other invention in human history."[3]

For this book linking mobile communication and the Internet, I think it is prudent to delay sharing any triumphant figures, for as the first few chapters will illustrate, there is not really a single mobile Internet to depict on such a graph. On the one hand, there is still only one Internet—the great network of networks[4] shuttling bits around the world. This Internet has an ever-growing array of both mobile and fixed paths to and within it. On the other hand, there is an increasing array of *means* of accessing and using the Internet, involving different combinations of hardware, networks, software, protocols, services, and content; and myriad *meanings* to that access, depending on users, skills, context, and cultures. These means and meanings, now increasingly mobile, yield innumerable different Internet experiences.

Yet across the world, there are some common properties of this more mobile Internet, sufficiently pervasive and essential to a range of human experiences as to merit not only enthusiasm,[5] but also careful reflection. My overarching goals for this book are to offer some reflections of this kind, and to use these reflections to explore the implications of the shift to a more mobile Internet for socioeconomic development and digital inclusion. Chapter 4 encapsulates these twin goals as the "After Access Lens" on mobile Internet use, which I hope readers will find to be a valuable resource for their own work or perspectives.

A short book about a massive topic demands considerable framing and narrowing. By situating this book project among my own affiliations, experiences, and academic communities, the rest of this chapter can identify opportunities and potential weaknesses in my endeavor, and should give the reader a sense of my approach to the topics I will be exploring.

Affiliations

Over two decades I have worked at the intersection of theory and practice concerning the use of information and communication technologies (ICTs) in the developing world. Currently I am senior director for research for Caribou Digital, a consultancy focused on building inclusive digital economies around the world.

Prior to joining Caribou Digital, from 2005 to 2014, I was a researcher at Microsoft Research (MSR), with instructions to do objective, long-term analysis to advance the state of the art in computer science and related fields; to help the company build better products and services; and to support the broader research community.[6] Thus while the book offers an analytical synthesis of a global trend, it does not do so from the perspective of a business strategist or product planner. It does not draw on any Microsoft internal documents, nor does it dwell on Microsoft's particular vision for the future of mobile computing worldwide. Instead, the bulk of the evidence presented in this book comes from public scholarship and practice.

Within MSR, I was a member of the Technology for Emerging Markets Group, a multidisciplinary team of researchers focused on the intersection between ICTs and socioeconomic development. Prior to MSR, I was a strategy consultant with Monitor Company, focused on regional economic development, and a postdoctoral research fellow at the Earth Institute at Columbia University, working on mobile health and mobile livelihoods projects in Rwanda.

My colleagues working in all of these organizations are talented, passionate individuals whose mandates are not simply to analyze the world, but to try to improve it through technical and social innovation, analysis, and practice. Our conversations and aspirations tended to take on a flavor of progressive, technologically optimistic interventionism. Elements of my own experiences at each of these institutions are probably visible in the perspectives and analyses I employ in this book.

Experiences

Another way to frame this book is to situate it as the extension of my own experience. In the chapters that follow, I strive to present and analyze a global phenomenon, emphasizing commonalities across mobile Internet technologies used every day by hundreds of millions of people in every corner of the planet. As I have mentioned, in one sense, there is only one global Internet. That is why, sitting in Cape Town, I can send an email around the corner, to Boston, or to Khartoum with negligible differences in difficulty or cost. However, in another sense, there are multitudes of Internets, some more mobile than others, appropriated, shaped, and reinvented by their users in an innumerable variety of contexts and cultures.

This second sense explains why my emails are most likely to go around the corner in Cape Town or back to my old home in Boston, rather than to Khartoum or to the rest of the places in the world I have not visited, or to people I have never met.

My experiences have brought me close to many elements of a more mobile Internet. I have conducted research on mobile phone use in the Global South for over a decade, including fieldwork in Rwanda, India, the Philippines, and South Africa. I have published studies on "development" topics like mobile health, mobile banking, education, livelihoods, and civic participation, as well as studies on ICT use in everyday life, from phone sharing and family use, to missed calls and bandwidth caps.

In particular, I will draw on four studies of mobile Internet use, conducted with colleagues between 2008 and 2013. Shikoh Gitau, Gary Marsden, and I studied "mobile only" Internet use among mobile phone users in Cape Town.[7] Our work on that project identified many of the constraints I bring up in part III, and yielded a paper that shares the name of the book itself, *After Access*.[8] Also in Cape Town, Marion Walton and I studied the interplay of shared access computer use and private mobile Internet use among teenagers.[9] That study helped me begin to think about the concept of digital repertoires, which anchors part III. Andrew Maunder and I did a study reflecting on Maunder's own startup, kuza.com, a portal for microentrepreneurs in South Africa and Kenya.[10] From my conversations with Maunder, I developed ideas around the relationship between the device and remote services that I build upon in part II. Finally, I worked with Preeti Mudliar and Bill Thies at MSR in Bangalore on a study of CGNET Swara, a platform for citizen journalism in India.[11] This work, exploring the intersection of "interactive voice response" (IVR) systems and basic phones, helped me question just what constituted Internet access, Internet use, and Internet experiences. Readers will see these ideas resurface throughout the book, from probing the edges of a mobile Internet experience (chapter 2), through effective use (chapter 8), production (chapter 10), and inclusive ecosystems (chapter 11).

Yet despite these projects and publications, my perspectives on mobile-centric Internet use around the world remain influenced by the experiences of where I have been, and remain limited by where I have not. Thus, I try to create work that integrates writings and perspectives beyond my own. Perhaps most germane to this book, I conducted a review of the literature on mobile telephone use in the Global South through 2008; that study, published in the journal *The Information Society*,[12] hints at some of the same integrative approaches I will employ here.

Communities

A final way to situate me (and yourselves, as potential readers) is to see this work in relation to the academic communities in which I participate. Although my doctoral training is in communication research, most of my scholarship has taken place in multidisciplinary communities rather than within my original disciplinary silo. Indeed, while I draw on research from a variety of disciplines—including communication, informatics, economics, sociology, anthropology, public health, design, and computer science—the book as a whole does not fit entirely within any of them. Instead, I can offer a small list of multidisciplinary communities from which this book draws, and to which it hopes to offer new insights in return.

Two of these—ICT4D and mobile communication studies—are the "primary" communities for this book, in that I hope this work will contribute to their core conversations and theoretical frameworks. In essence, this work is an extension of my ongoing efforts over more than a decade to cross-pollinate ideas and perspectives between these two groups.

In addition, I make connections to several other large, "extended" communities, including technology and society; development studies; social enterprise, design, and innovation; and new media/Internet studies. In these cases, though I do not intend to have the book directly challenge the frameworks and theories at their cores, I nevertheless think the material in the book may be a helpful input to conversations in each of them. I must admit that my engagement and review of these literatures is less comprehensive than in the case of the primary communities, so in these cases, I apologize to scholars who may not find their own works in the pages that follow. The literatures (and possible theoretical frameworks) available across multiple multidisciplinary communities are simply too voluminous to integrate into a single volume.

ICT4D

Throughout the book, I will draw on work by the community of researchers and practitioners exploring information and communication technologies for development (ICT4D or ICTD).[13] ICT4D, as I will call it from here on out, is multidisciplinary, attracting researchers and practitioners from engineering and computer science, social sciences such as economics and development studies, design, and critical theory. ICT4D conferences can be particularly lively—or frustrating—as participants wrestle with definitional complexities of everything from the nature of development to the technical algorithms underlying networking protocols. Although ICT4D echoes older conversations about communication, development, and the modernization paradigm,[14] many current perspectives in ICT4D stress intervention, innovation, and programmatic engagement with resource-constrained communities throughout the developing world. Others offer analyses of how people use, appropriate, and deploy existing technologies as part of personal, community, and national change processes.

At its best, ICT4D captures the complexities and potentialities of technologies, as applied and appropriated for a myriad of different uses under the complex and often contested banners of development, justice, and progress. At its worst, as critics point out, it can reflect shallow, Western, neoliberal, technologically deterministic approaches.[15] Yet on balance, after an initial burst of rose-tinted enthusiasm at the turn of the millennium, ICT4D has engaged in reflection and refinement of its theory, methods, and practice.

The mobile explosion has transformed the ICT4D conversation, and there has been a growing array of activity under the banner of "mobiles for development" (M4D), including practitioner communities,[16] and an academic conference series starting in 2008.[17] The World Bank made M4D the focus of its high-profile annual ICT4D report in 2012.[18] Individual mobile network operators, like Vodafone, and the GSMA (Groupe Speciale Mobile Association, or GSM Association, representing hundreds of mobile network operators around the world) have been keen to document and promote the economic and developmental contributions of the "Global System for Mobile Communications" (the dominant protocol for cellular wireless communication).[19] Pilot projects and specialized initiatives abound, and the GSMA boasts of over 800 M4D projects in its online database.[20] NGOs, universities, donors, and social enterprises are working to create specific mobile products and services. The M4D conversation has become multisectorial and international.

I have written specifically about M4D in the past,[21] and references to M4D studies of both the technical and social-scientific variety will appear frequently in the pages that follow. However, this book is not a *comprehensive* M4D text. More specific discussions of the exploding potential of mobile phones for use in traditional development verticals like, for example, in health,[22] agriculture,[23] learning,[24] disaster response,[25] governance,[26] civic engagement,[27] water, sanitation, and hygiene delivery,[28] and livelihoods[29] are available elsewhere. Indeed and instead, the book will help make a case that as mobile technologies increasingly touch the Internet, "M4D" as a term for a standalone field or even community of practice makes less sense.

Mobile Communication Studies

My other scholarly home is in the multidisciplinary conversation about mobile communication and society. Though this conversation sometimes draws on classic studies about the sociology of the telephone,[30] most of its research has emerged only after the turn of the millennium. Like ICT4D, mobile communication research has a journal, *Mobile Media and Communication*,[31] and a growing community of scholars and canonical texts.[32] This multidisciplinary conversation provides insights into what is unique about the mobile communication experience, and I will reference its works throughout the chapters that follow.

Even when not specifically about the Internet, the expanding literature on mobiles in the developing world has been useful. My 2008 review identified over 200 scholarly papers on the topic in 2008[33]; by now, the available literature has probably tripled. To wit: an issue of the interdisciplinary journal *New Media & Society* was called "Mobile Communication in the Global South,"[34] and drew on elements of Internet studies, mobile communication studies, ICT4D, and area studies. Indeed, mobile communication has become such a fixture of life in societies around the world that it is the subject of country-specific books, such as Horst and Miller's *The Cell Phone*, situated in Jamaica,[35] and Doron and Jeffrey's *Great Indian Phone Book*.[36] These comprehensive accounts capture the nuances and complexities of the intersections between mobile communication and society, which complement whatever cross-national, cross-contextual theorization I might accomplish here.

If my goal with ICT4D is to push for further refinement of its theorization and practice around mobile technologies (leading, perhaps, to the reintegration of M4D into ICT4D), then my goal for mobile communication studies is to update its frames toward mobile use in the Global South to reflect *Internet* use.

Extended Communities

Of course, my scholarly homes in ICT4D and mobile communication studies are not the only sources for analysis of the interactions between technologies and people, particularly in the Global South. For example, a thriving discussion around "community informatics" (CI) focuses on the intersections between digital technologies and strong communities,[37] but without the North-South and "development" frames often implied by ICT4D. Indeed, conversations about "technology and society," "science, technology and society" and "the Information Society" have been underway for well over a century.[38] Even an illustrative list would be too long. Works by Beniger,[39] Castells,[40] Wellman,[41] and Latour[42] are among the myriad lenses available to explore how technologies—lately and perhaps particularly information and communication technologies—are both reflections of and influences on daily life and the structure of societies.

Some conversations in development studies[43] are conceptually proximate to these information society questions. Lately, urgent, broad questions of globalization, inequality, and the spread of the information society demand a range of analytical frames from the pragmatic to the profoundly critical, of which ICT4D is only one sub-community. Manuel Castells' work, in particular, has influenced my own perspectives on mobile use in the developing world.[44]

Another community of practice has emerged from the business, social enterprise, and design communities, using the frames of *users*, *markets*, and *consumers*. With this reframe, enterprises and policymakers alike have become more aware of the potential application of "market-based solutions" to problems in development, and of the promise of "frugal"[45] or "inclusive"[46] innovation. Most famous, and perhaps most controversial,[47] of these frames is the notion of a "fortune at the bottom of the [income] pyramid"[48] awaiting companies that perfect the art of serving poor consumers. More broadly, however, many ideas around appropriate business models and design insights for resource-constrained communities have survived initial highs and lows in the hype-cycle, and have been embraced by a growing community of practice.[49]

The broadest and most heterogeneous of conversations on which I draw may be Internet and new media studies[50]. A thriving community of scholars has emerged, offering tens of thousands of assessments[51] of the Internet's birth, societal significance, and future trajectories. Importantly, there are transdisciplinary spaces, such as the ubiquitous computing and human–computer interaction (HCI) communities[52] and the annual conference of the Association of Internet Researchers (AoIR), that reflect and sustain the enthusiasm, nuance, rigor, and maturity of a conversation about "the Internet" no longer constrained to the PC. My discussion of a more mobile Internet, as it manifests particularly in the Global South, may be useful to this conversation.

However, permit me to make this point emphatically: this is not another book on all matters of the Internet in general, or even on the development impacts of the Internet; the topic is simply too vast. I am not able to describe the litany of ways in which Internet access and use may be unfolding in the Global South; instead, I will focus more narrowly on how some of those myriad ways are different *because the Internet is becoming more mobile*. Put another way, despite a robust array of literature about the Internet, now decades old, the user base is doubling (again), and this time our notions of what the Internet *is* will have to adjust as more of its users, particularly in the developing world, experience it primarily through mobile channels.[53] As communication researcher Harmeet Sawhney puts it, a shift to a more mobile Internet alters the "character of the Internet" itself.[54]

Another goal of the book is to complicate and update the established ideas in the popular and scholarly discourses about terms like "the mobile phone" and "the Internet." These have developed parallel literatures over the years, but the shift to a more mobile Internet challenges these distinctions. We are entering a period dominated by interlocking devices, services, and networks, yet increasingly providing heterogeneous Internet experiences. Nor is it sufficient to skip completely to a framing of "convergence" where all devices deliver comprehensive and equivalent experiences. It will take a few chapters to get there, but this book argues that research studies, policy pronouncements, and development interventions that turn on vague concepts of "the mobile phone" or "the mobile Internet" will be inadequate to account for the changing communicative landscape.

Intersections and Antecedents

These various conversations and communities of practice are not sharply delimited; I presented them sequentially mostly as a way to signal the multidisciplinary breadth of research and practice informing these topics. Indeed, cross-pollination among these conversations is common, and several authors working at the *intersections* of these conversations have particularly influenced my research and practice over the years. Here are two illustrative examples. Jack Qiu's work on China's "working-class network society" and "the information have-less," centered largely on prepay phones and text messaging (SMS), offers a lens on "informational stratification" rather than classic digital divides.[55] Researchers at the research group LIRNEasia in Sri Lanka have been promoting research on "more than voice" mobile services for the poor, and have been enthusiastic champions for a perspective that looks beyond the desktop PC as the digital technology best suited to help resource-constrained people participate in the information society.[56]

The case of South Africa has been particularly instructive for me as a researcher, not only because it is the place where I lived while writing this book, but also because it is an early adopter of mobile Internet in the Global South and is simultaneously home to many researchers and practitioners working on these issues. These include Wallace Chigona's writing about the mobile Internet and "social inclusion in developing countries,"[57] Peter Benjamin's work on mobile health,[58] the late Gary Marsden's work on mobile interfaces and human–computer interaction,[59] and Marion Walton's work with teenagers and mobile literacy.[60] I will mention several others throughout the book.

Of similar central importance to this book are those projects that wrestle specifically with the ideas of a mobile-centric or mobile-only Internet experience in the Global South. Katy Pearce[61] at the University of Washington and Nimmi Rangaswamy[62] have been particularly prolific early voices on this matter.

With so many communities in the mix, there is a lot of ground to cover. It is my hope that the total amount of multidisciplinary scholarship on a more mobile Internet for development and digital inclusion is still limited enough that I can represent much of it, but readers should not assume that the work represents certainties or consensuses where none (yet) exist. The shift to a more mobile Internet is a multifaceted phenomenon, and it is happening quickly, almost everywhere at once. That said, I am grateful for what I am able to read from and discuss with peers in the communities I outlined earlier, and feel privileged to be able to offer this contribution to research and theory on an increasingly mobile Internet.

Approaches

It should be evident from the breadth of the academic communities involved that there is unlikely to be agreement among these peers about the level at which to engage with (or alter) the deep structural linkages among technology, society, and development. I am wary of disappearing into epistemological rabbit holes at this point; to do so would take the book in a different direction. Don Slater's discussion of "development narratives"[63] in his book *New Media, Development & Globalization* is a much better resource for reflections of this kind. Perhaps so too are discussions about the durable, unavoidable cleavages between administrative (empirical) and critical research, recognized and vigorously debated in the field of communication for decades now, but by no means exclusive to it.[64]

Let me instead offer an extremely simplified taxonomy wrapped in a biological analogy, and stripped of references for simplicity's sake. Some ICT4D "development narratives" suggest ICT use can solve problems, as a medicine treats a disease in an unwell body. Others suggest ICT use merely reflects and channels deeper problems and systems in the body, more like a symptom than a problem. Others go deeper still, arguing that we must understand that ICTs have come to be part of the body itself—ICTs seen as circulatory, neurological, skeletal, or cognitive systems in the organism. Finally, some offer critiques of the body's current ICT systems as profoundly flawed, suggesting that differences in ICT use are a problem themselves, replicating structural inequity and injustice. After more than a decade in the ICT4D field, I expect to hear new articulations of each of these perspectives (and several more), with freshly divergent (or absent) framings of the problem and the desired "development outcome" at each new conference or roundtable I attend.

I am not frustrated by these durable tensions, however; nor do I think a total epistemological victory of one camp over all the others is anywhere within sight. Indeed some of the most interesting work in ICT4D is coming from participating scholars whose work interrogates the very idea that information technologies can be "for" "development" at all. Any comprehensive treatment of a phenomenon like the shift to a more mobile Internet demands attention to critical views that see mediated power and structure as a problem as much as a potentiality. However that does not mean I am not impressed and humbled by the innovative technologies and "solutions" that those with the interventionist, design, technical, and administrative perspectives seem more likely to create.

Thus, I will endeavor to keep any breathless, context-free cheerleading for the mobile telephone and the Internet in check. I do not believe that either cluster of technologies is capable of solving all the problems of inequity, strife, and poverty on this crowded planet. However, just because ICTs are not the exclusive solution to humanity's problems does not mean they do not structure daily lives and livelihoods around the world, and certainly does not mean they are immutable forces of nature. Technologies, even ones as vast and complex as the Internet and the mobile networks, are the aggregation of thousands of design, investment, and policy decisions made by people and organizations in interaction and collaboration with their users, markets, and constituencies. Later chapters will offer suggestions on how we might improve the decisions we make about mobile devices, services, technologies, and policies, in ways which favor digital inclusion.

In so doing, however, I must acknowledge that my approach is mildly technologically deterministic; perhaps Jo Tacchi's framing of "contextualized affordances"[65] or Sun's ideas of relational affordances[66] can characterize my approach to the estuarial space between determinism and appropriation. I will introduce caveats and reminders throughout this text about how no technology can be fully understood (nor any development intervention be truly successful) without understanding the particular human context and relationships in which it is situated. However, this work builds on the idea that there are, still, properties and potentialities that carry from setting to setting. I will focus on how, *caeteris paribus* (with other things the same), different configurations of digital devices create different opportunities for people to manipulate and produce digital information, and therefore to coordinate, propagate, and produce the stuff of daily life and of development. The book posits both that ICTs have an impact on development, and that on balance, ICTs can be shaped by innovation, intervention, and policy to be a force for progress, productivity, and prosperity rather than inequity, domination, and despair.[67] It is hard, some days, to stick to these propositions, but I am working under the general frame that things are still "getting better."[68]

The themes I explore in the book do have implications that are broader than the frames of "development" and "developing countries." I use the dichotomies Global South/North and Developing/Developed World somewhat interchangeably throughout the book. Each term has its supporters and detractors, but I use each of them to highlight and reflect on communities or contexts (at local, regional, national, and global levels) where resources and network power are scarce, constrained, or unequal. Yet there are patterns of inclusion and exclusion, centrality and marginalization, abundance and constraint in nearly every community on Earth[69] and the shift toward a more mobile Internet is underway in almost all of them. Thus the term Global South does a bit better than Developing World in capturing the granularity and universality of the dynamics of power and poverty.

The intended audience for the book is rather broad. I am, of course, writing for scholars in all the communities I listed earlier, interested in Internet use in the developing world and the Global South, as well for those interested in matters of digital participation and inclusion in the information society. However, in addition, I am writing for policymakers trying to understand, reframe, and improve mobile Internet technologies, infrastructures, and services in the developing world, and I am writing for designers, technologists, and entrepreneurs seeking to harness mobile Internet technologies to create new and more useful products and services, and for development practitioners hoping to apply such technologies "for development."

Thanks to many years working outside the traditional confines of academics, I have friends and long-time colleagues fitting each of these audience categories. I have tried to fashion a book that is approachable enough to be useful across these different sets of demanding readers. Though it is not an easy task, the result is the book I wanted to write, and one I think this particularly multifaceted topic demands.

The Book from Here

Across these audiences, therefore, the book might influence how some readers think about the so-called digital divide, or might influence how others think about (re)framing the development project in the changing context of the emerging global information society. However, these are not the primary intended outcomes. Rather, and much more narrowly, this book synthesizes current research and theory from two distinct communities I have mentioned above—ICT4D and mobile communication studies—to examine the shift to a more mobile Internet and its implications for the trajectory (and pursuit) of socioeconomic development and digital inclusion. What, exactly, your model of development might be remains up to you.

Comprising this introduction and the next three chapters, part I focuses on defining "mobile Internet." Chapter 2 details how mobile Internet technologies are a work in progress, with rapid advances in coverage, affordability, and functionality. Chapter 3 continues the task of re-definition with two specific, complicating questions: "how mobile?" and "which Internet?" It suggests that the gradations in experiences between users are so great as to complicate any bifurcation into user and nonuser. Thus, instead of offering a specific definition of a discernable, standalone, separate mobile Internet, chapter 4 identifies six general properties that tend to differentiate mobile modes of Internet use from "fixed" modes. The chapter suggests that each has played a role in explaining the rapid growth in mobile Internet use, and uses these properties to introduce and anchor the "After Access Lens," which will be used throughout the book.

Part II merges mobile communication research with ICT4D to identify two broad new potentialities afforded by mobile Internet technologies. Chapter 5 starts the section by contrasting research on the social and economic impacts of landlines, mobile phones, and the conventional Internet. It isolates cases where a more mobile Internet allows different interactions with data and services residing remotely and, in so doing, allows its users new ways to renegotiate their orientation to physical places. Further articulating the After Access Lens, it argues that mobile modes of Internet use afford greater options to have place(less) informational interactions, and paradoxically, to engage in more place(full) ones as well. Chapters 6 and 7 explore these new behaviors in the context of ICT4D.

Part III concludes the articulation of the After Access Lens, focusing on how the same properties that have fueled the boom in access also present considerable constraints to engaged, effective Internet use. Chapter 8 merges the two key concepts for use in part III: (1) from mobile communication research, *digital repertoires* (which allows me to distinguish mobile-only from mobile-centric Internet use); and (2) from ICT4D, the idea of *effective use*, which is a flexible way to evaluate how individuals and communities might use ICTs (as tools) to achieve their development goals. Chapters 9 through 11 take the three major constraints in turn: a *metered mindset*, *limited production scenarios*, and *circumscribed structural roles*.

Part IV concludes with a single chapter wrapping up the implications of the full After Access Lens for ICT4D and mobile communication research, with nods to implications for the broader scholarly and practitioner communities in adjacent fields. In general, the chapter rearticulates how the shift to a more mobile Internet creates new opportunities and new challenges for ICT4D in assuring that (in the words of Michael Gurstein) "the internet is and continues to be a resource available, usable and of equitable benefit to all."[70]

2 Mobile Overtaking Fixed, Again

Whether 2002 was a long time ago, technologically speaking, is largely a matter of perspective; some may have difficulty recalling a digital landscape before Facebook, YouTube, and the iPhone. Regardless, the world reached an important telecommunications milestone that year. For the first time, there were more connections to the global telephone network from mobile devices than from fixed (landline) telephones—1.1 billion mobile connections in all.[1] In the twelve years since that point, mobile network operators (MNOs) around the world added five billion more mobile-cellular connections, particularly in the Global South, while fixed subscriptions stayed virtually flat at 1.1 billion.[2] Mobile telephony did not just overtake fixed telephony. Mobile ran fixed off the road, took its wallet and keys, and sped off into the sunset with 6+ billion connections.

Now, the telephony race has a sequel. Mobile Internet is overtaking fixed Internet. And, as was the case with voice telephony, the Global South has much to gain from this shift to a more mobile Internet. Yet identifying a similar single transition date will be more difficult than last time. This chapter begins by illustrating how the transition depends on what we count, acknowledging that whatever number one might pick is sure to be eclipsed. The second half describes how a broad array of stakeholders is still striving to use mobile technologies to make the Internet more affordable and accessible to those still without a connection.

Counting

If we wanted to identify the moment when mobile Internet overtook fixed, which metrics would we use? Even more than was the case with telephone lines, there is increasing divergence among connections, devices, users, and activities involved in connecting to the Internet via mobile technologies. I will return to the resulting heterogeneity and stratification of mobile Internet experiences in the next chapter. For the time being, in this section, I present some metrics that could be used to understand the growth in importance of mobile technologies as a means to access and use the Internet.

Connections

One way to represent the transition from fixed toward mobile Internet would be to count Internet *connections* (subscriptions), as the International Telecommunication Union (ITU) did with telephone lines. By this metric, the ITU's own estimates would suggest the fixed/mobile data inflection occurred as early as 2008, when mobile broadband subscriptions nudged ahead of fixed connections, 422 million to 411 million. Since then, the lion's share of the growth has been on the mobile side. By 2014, the ITU could count three times as many mobile broadband connections as fixed broadband subscriptions, 2.3 billion to 711 million.[3] Figure 2.1 represents this growth in subscriptions in per-capita terms, commonly called the "penetration" of subscriptions in a country. Researchers at LIRNEasia point out two significant problems with this approach. First, every new active subscriber identification module (SIM) card with the *capability* to access data is considered a subscription, whether or not its owner ever uses it to draw data.[4] Second, since people can own more than one subscription, in many places there are now more such mobile-cellular subscriptions than there are people (more than 100 lines per capita). In the developing world, reported penetration ratios are lower, but the patterns are the same, with a stunning rise in mobile broadband from 1 in 100, to 20 in 100 over the same seven years. Fixed broadband subscriptions in 2014 were still only 6 per 100. The telecommunications network manufacturer Ericsson releases an annual "Mobility Report"; they project global mobile broadband subscriptions to continue to grow 20 percent annually through 2020, resulting in over 8 billion mobile broadband subscriptions worldwide.[5]

Devices

One can also compare counts of *devices*. Even a conservative interpretation of sales figures, accounting for only smartphones, again suggests we have

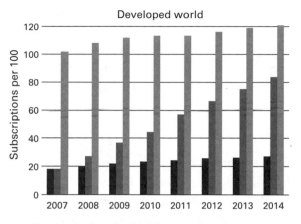

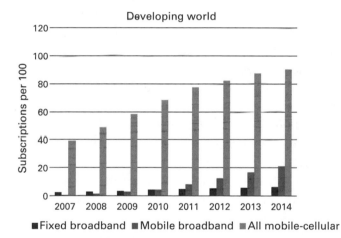

Figure 2.1
Active broadband subscriptions per capita. Source: ITU Key 2006–2014 ICT Data.
Note: broadband is >256kbit/s.

passed the mobile/fixed inflection point. Worldwide, unit shipments of new smartphones and PCs outpaced desktops and notebooks in late 2010, and industry analysts suggest that the total installed base (devices in use, no matter how old or new) of mobile Internet devices (smartphones plus tablets) eclipsed those of "fixed" Internet devices (laptops plus desktops) by mid-2013.[6]

Users

Still another way is to compare counts of *users*. This is a more difficult analytical task, since without costly and comprehensive surveys it is difficult to identify proportions of populations holding multiple devices or subscriptions, as well as to account for sharing in families, between friends, in workplaces, and in schools.[7] Indeed, the GSMA released a report in collaboration with the consultancy A. T. Kearney estimating that the six billion mobile subscriptions active in 2013 were concentrated in the hands of just over three billion unique individuals.[8]

The ITU estimates that at the end of 2014, there were 2.9 billion Internet *users* (fixed, mobile, or both) around the world, driven by 2.3 billion mobile broadband subscriptions and 711 million fixed broadband subscriptions.[9] These are drawn from national level estimates, however, and the exact tally of people accessing the Internet via mobile broadband, fixed broadband, or both is not part of the ITU dataset.

One could assemble a tally of well over a billion mobile Internet users from just two sources without relying on GSMA or ITU estimates. Facebook reports that by early 2014, more than a billion of its 1.3 billion users had used the service via mobile devices,[10] so that accounts for one billion. China is the world's largest Internet community with 632 million users in mid-2014; in a telephone (fixed and mobile) survey of 35,000 users, 85 percent reported being online via a mobile device, versus 81 percent reporting fixed use, the first time the official statistics from China Internet Network Information Center had mobile overtaking fixed.[11] That study suggested another 500 million unique mobile Internet users, living outside Facebook's footprint, as it is blocked in China.[12] By the users' metric, whether assembled by estimates or user logs, the inflection point has probably been reached, with over a billion and a half mobile Internet users so far, though the precise date at which mobile overtook fixed may never be known.

Activities

Consider figure 2.2. Statcounter[13] counts individual visits to over three million websites around the world, drawing information from the visits about the kind of device visitors are using. Over time, the proportion of visits coming from mobile devices is increasing. This trend is apparent worldwide, and by mid-2014, both South Africa and India had passed the inflection point. That is a massive shift toward mobile in five short years, and this analysis does not even account for non-browser-based Internet activity on mobile devices coming through apps.

Apps are a broad term for the enormous variety of downloadable software programs, from games and chat to word processors and medical reference texts, used to customize and expand the capabilities of mobile devices. I will say more about apps in later chapters. For now, they help support another activity-based metric illustrating the magnitude of the shift to mobile. Benedict Evans, of the Venture Capital firm Andreessen Horowitz, cites figures from the research firm comScore that estimate that in 2014, United States users spent more time (measured in billions of minutes) in mobile apps than they spent on the World Wide Web as a whole, across mobile and fixed platforms combined.[14]

An Alternative Frame Is to Look at Coverage

Thus, the exact date of the inflection when mobile Internet overtakes fixed Internet is difficult to ascertain, as definitional specificities intermingle with the complexities of measuring behaviors at a global scale. There may be no singular announcement of the kind the ITU issued in 2002 for telephony. Yet the general trend is unmistakable—the Internet is more mobile with every passing day.

In the meantime, and for a more durable metric indicating the potential of mobile Internet to surpass fixed Internet, it might be more helpful to look at *coverage*, which can be more reliably estimated for cellular/wireless technologies by matching tower locations and footprints to population censuses. Potential coverage, of course, does not translate to actual users, but gives a sense of who might be (or might know) a user.

By this metric, the current footprint of potential mobile Internet users is simply remarkable. The ITU estimates that by the end of 2012, 90 percent of the world's population, including, critically, 87 percent of its rural inhabitants, lived under a mobile/cellular signal.[15] Ericsson's mobility report

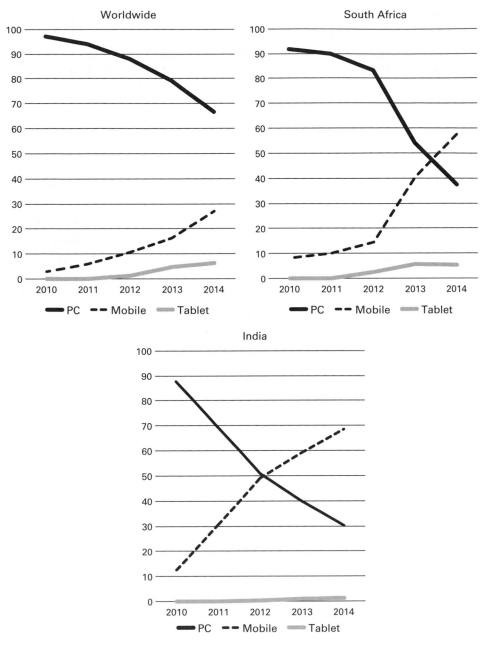

Figure 2.2
Visits to three million websites, by source. Source: gs.statcounter.com.

contains similar estimates. Their baseline 2013 figures put 85 percent of the world's population (people, not land area), including 70 percent of the population in the developing world, under areas covered by basic mobile services—voice communications, SMS, USSD, and, most important for this conversation, usually at least a slow second-generation (2G) data connection.[16] By the same estimates, 60 percent of the world's population lived under a third-generation (3G) broadband data connection like WCDMA or HSPA, while 20 percent lived under 4G—fourth-generation "Long Term Evolution" (LTE) signals—with speeds rivaling that of cabled broadband.[17] Ericsson projects that by 2020, a remarkable 90 percent of the world's population will have access to 3G broadband wireless connections, and 70 percent will have access to 4G connections.[18]

Expanding Access

Successes and progress notwithstanding, expansion in coverage is not complete, and the push toward eliminating the "access" challenges continues. Each metric around access (and use) of the Internet will continue to rise, in no small part because efforts are underway to further tailor mobile Internet technologies for greater adoption by billions of people near or below the poverty line. Let me stress that there is no victory to declare before the 15 percent living outside a mobile signal has practical access to *any* telephony, and before many more have reliable and affordable access to telecommunications networks. However, overall and over time, data-enabled mobile devices will become more *affordable*, mobile networks will become more *available*, and the tariffs to use them will become more *reasonable*.

There are extensive scholarly and policy discussions about how to update national-level connectivity initiatives for the era of mobile voice and mobile data. In this section, I will highlight a few elements of this literature, but would direct readers to other sources for more comprehensive reviews.[19] Instead, my aim is to illustrate some of the work still underway, both to squeeze more reach and performance out of existing mobile/cellular systems, and to promote and deploy alternative solutions. Some further improvements will come via regulation and policy interventions, some by new technical innovation, and some through creative application of existing technologies in new settings. Note there is not a specific, settled prescription that will work everywhere. Incentives, priorities, and perspectives will differ among policymakers, practitioners, and technologists, and particularly across national borders, contexts, and regulatory structures.

Frugal Hardware, Frugal Software

Organizations around the world are working to make data-enabled mobile devices less expensive to acquire, and less expensive to use. Engineers and computer scientists sometimes label the low-bandwidth, low-power challenges these devices must operate in as "constrained computing,"[20] which demand "frugal innovation"—invention and adjustment by and for people with little to spend on computing.[21]

Moreover, in many cases, there is little to spend. In 2008, the World Bank revisited their assessments of poverty and incomes around the world. In PPP (purchasing power parity) dollars, which tend to be higher in developing countries, compared to market exchange rate dollars: "Half the world's people consumed less than PPP $1,300 a year and the bottom quarter less than PPP $660 in 2005."[22] Populations, currencies, and income levels continue to change, but it seems reasonable to keep these thresholds in mind as we discuss the challenges of affordability. For any technology to expand below the world's income medians, it must be very inexpensive, indeed.

Lowering Up-front Costs for Users

In most of the world, phones are sold unlocked and at full retail cost, instead of subsidized or bundled into a monthly service contract as happens in the United States. Thus, for many people, the up-front cost of data-enabled mobile devices is a barrier to adoption. In early 2014, Dennis Woodside, CEO of Motorola, put it bluntly: "In much of the world $179 is a lot of money, so there's a big market at a price point of less than $179. We're going to look at that, and just delivering on that value promise is super important. I mean, why can't these devices be $50? There's no reason that can't happen."[23] He had better keep working, as low-cost smartphone manufacturers based in China (Xiaomi, ZTE, and Huawei) and India (MicroMax) are aiming for just that.[24]

Materials and manufacturing account for some of the costs of a device. Indeed, a modern handset draws on a remarkable combination of global supply chain activities,[25] including sourcing raw materials and pre-manufactured components like circuits, assembly by subcontractors, primary assembly, packaging, and distribution. Some costs cannot be reduced by better engineering or thrifty company practices. For example, many governments set import duties on handsets.[26]

Smartphones are fantastically complex devices with long heritages of innovation and investment behind them. Yet the resulting thicket of interconnected, highly contested patents and licensing agreements in the handset industry presents a significant cost for the production of data-enabled phones. A working paper by Armstrong, Mueller, and Syrett estimates the costs of patent licenses—the "royalty stack"—on a hypothetical $400 LTE smartphone at $120.[27] This is approximately the same as the cost of materials. Work underway by Nehaa Chaudhari at the Center for Internet and Society in Bangalore (Bengaluru), India, is exploring the implications of such a stack for affordable smartphones at the sub-$100 level, illustrating the ongoing negotiations (and litigations) among Indian and Chinese handset manufacturers, international license holders, and, possibly, the Indian government regulators to find licensing regimes that work for such low-cost devices.[28]

The license (or lack thereof) for the operating system (OS) is one particularly important input to the cost of a smartphone. Different manufacturers have different approaches.[29] Every phone by Apple runs its iOS operating system, and conversely Apple makes every device running iOS, so there is no license to pay. Google's Android, by far the top-selling smartphone operating system on the planet in 2014,[30] takes a different approach. Google leads the development of much of Android, but the operating system is open source and is available to handset manufacturers at no monetary cost. The absence of a substantial licensing fee has made Android appealing for low-end devices, and Google is using its "Android One" reference program to help low-cost manufacturers make better phones at the sub-$100 price point.[31] In 2014 Mozilla, makers of the open-source Firefox browser, launched a $25 smartphone running its low-cost Firefox OS instead of Android.[32]

However, for the 25 percent of the world living on $660 (PPP) or less per year, and even for the 50 percent living on $1,300 or less, even a $25 phone (in market-rate dollars) is a significant outlay. Thus, secondhand markets for handsets thrive in developing countries.[33] Repair matters as well; unbranded community shops can keep a handset working,[34] and play a key role in increasing the affordability and overall lifecycle of a data-enabled handset.

Governments can subsidize the cost of handsets for particularly vulnerable groups, as, for example, Indian and Nigerian politicians have suggested in the midst of elections.[35] The evidence of the efficacy of these direct subsidies is scarce. However, it is possible that as states achieve near-universal adult adoption of mobile devices, they will decide to help bring the most vulnerable, marginalized, and resource-constrained members of their societies online regardless of the handset cost or demonstrable economic return on investment.

Practitioners and technologists should also keep sharing scenarios in mind. It is tempting to think of the mobile handset as belonging to— and paid for by—a single individual, but in resource-constrained settings, handsets are often shared.[36] Sharing happens in villages,[37] among family and friends,[38] and more formally through programs like Grameen Bank's Village Phone, in which a microentrepreneur owns and operates a mobile phone as a public payphone for her community.[39] Grameen has updated its Village Phone program for the age of mobile data, deploying "Community Knowledge Workers" with shared smartphones across Uganda.[40] Other models of smartphone sharing will emerge in other settings as long as the demand is there among those without handsets of their own.

Lowering Running Costs: Compress, Conserve, Transpose, Cache

Yet as a car needs gas, a data-enabled handset needs data. Reducing the ongoing costs of mobile Internet use is the other side of the coin to reducing up-front costs. Mobile data remains expensive, relative to fixed-line delivery, so just as resource-constrained drivers like fuel-efficient cars (perhaps better yet, motorcycles), resource-constrained mobile users like data-efficient devices. Unfortunately, much of the Internet itself is changing, getting "heavier." The average web page in 2012 may have been more than twenty times as big as the average page in 1999.[41] Video is particularly bandwidth-intensive, and is increasingly common online.[42] For resource-constrained users more likely to be using low-end handsets to access relatively slower networks, ongoing efforts to make the mobile Internet "lighter" are of paramount importance.

Some lightness comes simply from careful design and deployment. The World Wide Web Consortium (W3C) champions best practices for the mobile web[43] and for mobile web applications,[44] reminding app creators that "the most effective way to ensure that applications run smoothly and with low latency is to minimize use of device memory, processor power, and network bandwidth which are more limited on mobile devices than on the desktop." Concisely, less is more.

Some services appeal specifically to user demands for frugality. For example, BlackBerry has a loyal cadre of middle-income users in many emerging economies.[45] Two of its continued strengths are the BlackBerry Messenger service, BBM, and its easy-to-understand data plans.[46] Elsewhere, messaging services like WhatsApp, Viber, Line, and WeChat are finding similar traction with data-enabled phone users in emerging markets who might otherwise send expensive SMS messages.[47]

Beyond lean websites and efficient apps, additional frugal approaches include compression and optimization of Internet data for mobile devices.[48] For example, instead of accessing the World Wide Web directly and rendering pages on the mobile screen, some mobile browsers elect instead for server transcoding, which routes traffic first to a proxy server. Before the proxy server sends data on to the handset, it compresses or transcodes (transforms) the data to reduce its size, optimizing for lower-bandwidth mobile scenarios.[49] The best known of these browsers is probably Opera Mini, designed specifically for lower-end phones. Opera suggests that at least 250 million people used the Opera Mini compression browser in January 2014, enjoying data compression rates of up to 90 percent.[50] Beyond the browser, in 2014, Opera brought transcoding to video and apps via the "Max" application for Android.[51] Binu offered a downloadable app (with compression) for entry-level phones that allows users to browse the web and access Facebook and Twitter.[52]

Extend Networks (at Reasonable Tariffs)

Alas, the most affordable and data-efficient devices in the world will not help a user connect to the Internet if there is no network connection available. Service providers continue to deliver greater network connectivity (via the mobile/cellular network) in more places at more affordable rates. I put coverage and affordability together since they are often interconnected. In general, sparsely populated rural areas remain hard to serve economically with conventional mobile/cellular approaches. I will get to satellites and other alternative networks below; for now let's focus on mobile towers and mobile wireless spectrum.

One fixture of mobile connectivity is that a great deal of it happens across "limited" bands of electromagnetic spectrum controlled by national regulators and auctioned to providers. Thus mobile (cellular) Internet prices are determined not only by what it is technically possible to supply, but also by the tapestry of national and international regulatory and standards-making agencies present in each country. Although the details of the inter-play between network technologies and network regulation are largely beyond the scope of the book,[53] a few words are in order. In theory, those regulating the mobile Internet might draw on desirable guiding principles of previous technologies—universal service from fixed telephony, optimal spectrum allocation from cellular voice, and the principles of net neutrality and openness from the Internet. In practice, the management of different layers and classes of services is a messy affair, mixing what is best for voice and data.[54] The capabilities and strategic priorities of mobile network opera-tors, equipment manufacturers, exchanges, backhaul data carriers (those who manage the international and long-distance cables), satellite provid-ers, incumbent telecommunications companies, Internet service providers (ISPs), and Internet-based services are not always likely to be aligned.[55] That said, at the broadest level, independent, engaged regulators pursuing liber-alization strategies and competitive markets might still be the most produc-tive path.[56] One general principle is Samarajiva's idea of a "Budget Telecom Network Model," where pay-as-you-go pricing and healthy competition allow companies to serve customers with very low monthly spending allow-ances.[57] This section lays out four overlapping efforts aimed at expanding affordable mobile service to more people: expanding geographic coverage, improving network performance, improving spectrum availability, and establishing reasonable tariffs.

Expanding Geographic Coverage

Those living in areas without mobile signals would probably be pleased if mobile network operators could be incentivized, or simply required to cover more terrain. Barring a change in the availability and costs of satellite or landline Internet, practical access to the Internet remains closely tied to access to a mobile telephone base station. If the patterns of rollout and adoption of mobile data follow those of mobile voice, service provision to urban areas can be satisfied through liberal, competitive mar-ket mechanisms.[58]

Yet still, not everyone lives under a signal. Whether it is 5 percent, 15 percent, or 25 percent of the total world population, those without access to wireless data are overwhelmingly rural, low resourced, and relatively harder to serve,[59] facing what Dymond and Oestmann[60] describe as the intertwined challenges of poverty and isolation. These analysts distinguish between a "market efficiency gap" and a broader "access gap," where the latter "recognizes that additional intervention may be required to reach areas and groups that policymakers legitimately wish to reach . . . but who will not be served even with the most attractive and liberalized market conditions."[61] These are complex policy decisions, and the right to data access is not yet settled,[62] yet as is the case with voice, these hard-to-serve areas and populations may require creative and aggressive interventions in order to encourage, if not guarantee, affordable service to rural users.[63]

Improving Network Performance

MNOs are building new towers and upgrading existing ones, replacing 2.5G with 3G, 3G with 4G, and so on. Beyond these upgrades, however, there are other steps network providers can take to improve network speed, reliability, and capacity. Improving the reach of fiber backhaul inside countries and expanding connections to international cables is critical in lowering the cost and increasing the availability of wholesale data to ISPs and MNOs.[64] In addition, expanding peering arrangements (the agreement between bulk carriers to exchange data across networks without billing each other), and building physical Internet exchanges (linking networks) and data centers (housing servers)[65] in developing countries can dramatically reduce the time it takes Internet traffic to find and return from major international resources.[66]

Other approaches can treat different traffic with different priorities. Packet shaping—a commonly used tool in businesses and within ISPs—ensures that bytes deemed more time sensitive by the provider find their way to and from users more quickly than those bytes considered less important. ISPs may elect to "throttle," "shape" the Internet connections of some users, "cap" the amount of usage by others, or a combination of these.[67]

Yet as I just indicated, these technically possible optimization choices raise policy implications for what is politically and economically acceptable and desirable. During the development of the fixed-line Internet, many policymakers and regulators worked to build and defend a principle of net neutrality in which carriers and ISPs did not discriminate between types of traffic; those discussions and conflicts continue today.[68] Yet, in the mobile domain, as bandwidth constraint becomes so much more palpable, operators are looking for ways to shape and prioritize traffic even more aggressively than some fixed-line Internet operators might seek to do. Some of these goals are intended to create a more efficient network, others are not so well intended: for example, banning voiceover IP in a territory in order to defend the revenues of incumbent telephone companies.[69] I will return to another specific case, the subsidization of selected content via "zero rating," in much more detail in chapters 9 and 11.

Improving Spectrum Availability

Other advances are being made in spectrum availability—making sure that existing bands reserved for mobile/cellular operators are efficiently priced, managed, sold, and utilized.[70] In addition, efforts are underway to provision new "unused" frequencies from other parts of the electromagnetic spectrum. Both of these general efforts improve the supply of bandwidth available to users.

The process of delivering data across the *existing* spectrum also used for voice remains a moving target, with new efficient technologies like LTE replacing older protocols. In that case, some reconfiguration of licenses or trading of spectrum between license holders, or both, may be in order.[71] What made technical sense in 2005 may make less sense a decade later.

In terms of *new* spectrum, to ease the coming crunch, many operators have eyes on the "TV whitespaces"—frequencies left vacant as countries shift from analog broadcast TV to digital signals.[72] As it happens, TV whitespaces spectrum can be good for serving rural areas across long distances. Microsoft and Google are both supporting TV whitespaces trials in Africa.[73]

Establishing Reasonable Tariffs

To mobile network operators, choices about pricing are closely tied to investment in infrastructure and spectrum licenses. Pricing decisions and promotions are particularly complex in an environment with rapidly rising demand for data services, flat demand for voice, and even falling demand for SMS.

Critics suggest that in many countries, tariffs for mobile voice (and data) may simply be too high, and that poor households' proportional

expenditure on airtime represents a failure in public policy.[74] I do not pre-
tend to know the "fair" cost (if any) of a minute of airtime, or an SMS, or
specifically the cost of a byte of data. However, in 2011 the UN Broadband
Commission suggested a target of less than 5 percent of gross national
income per capita. Most of the developed world had met that target by
2013; but the commission noted in 2014 that "broadband still remains
unaffordable in many parts of the developing world."[75] I will return to pricing issues in more detail in chapter 9.

A final note on tariffs and pricing: thanks to batteries, and unlike most
desktop PCs, some mobile devices can operate in places that are off the
electrical grid, but electricity is a huge issue, and in places where power is
scarce, mobile Internet use is considerably more difficult.[76] Jan Chipchase
suggests that in some places, like rural Myanmar, the cost of the charge to
watch a movie on a phone can outpace the cost of the data![77]

Introduce Alternative Networks

This section has focused mainly on efforts to expand affordable "archetypal" mobile Internet experiences—using handheld devices, running a
mobile operating system, connecting via the cellular network. However,
refining the existing system is not the only possible path. Alternative networks will be part of the tapestry of solutions that will help extend Internet
access to geographies that do not have cellular coverage, and will expand
options within areas that already do.

WLANs, Hotspots, and Community Wi-Fi

Wi-Fi networks, provided by telecenters, schools, libraries, community hotspots, and wireless local area networks (WLANs), will play an important
role in easing pressure on cellular networks, and on increasing the quality
of connection available to users.[78] After all, many smartphone users in
developed countries prefer Wi-Fi for much of the heavy downloading and
uploading they do during the day.[79] The same should be true for users in the
developing world,[80] but Wi-Fi availability varies considerably and is particularly scarce in the same rural areas where cell coverage is poor. Steve Song,
a leader in the ICT4D community, blogged about Wi-Fi in 2012:

the UN Broadband Commission's recently published report on Achieving Digital
Inclusion mentions Wi-Fi exactly twice, both times parenthetically. Mobile operators would like you to believe that the future of mobile broadband lies in the LTE
networks that they are building. And certainly that is partly true but only partly. If

the Mobidia stats [suggesting 70 percent of mobile data in the US gets offloaded to Wi-Fi] are to be believed, about 30 percent true . . .

. . . In any discussion about the mobile broadband future of Africa, Wi-Fi is simply not part of the discussion. Yet the evidence is before our eyes of the strategic importance of Wi-Fi to our "mobile" devices. It's cheap and fast and grew to solve the problem of affordable access by chance not by design. It happened because Wi-Fi is an open space for technology developers to innovate. No carrier agreements required.[81]

Despite the UN Broadband Commission's inattention, some business models and pilot projects are exploring Wi-Fi as the primary connection for low-resource users in the developing world, including via community hotspot resellers[82] and LTE-to-Wi-Fi conversion hotspots.[83] In Africa, Airtel is building out its own Wi-Fi network in order to ease demand on its cellular towers.[84] Meanwhile, Song's current project is the "Mesh Potato,"[85] a wireless mesh network based on Wi-Fi that is designed to extend voice and data connectivity to communities piece by piece, growing as the demand grows.

Speaking of communities, it is an unfortunate quirk of the network that while not every byte needs to go up to a national (or international) server to be useful, most bytes do just that. For example, computer scientist David Johnson and his colleagues found a high degree of local-to-local digital interaction in a village sharing a WLAN in Zambia; much of the bandwidth consumed by people on the local network was devoted to social networking between local, proximate users.[86] In current mobile pricing regimes, sending a message to a friend tends to cost the same whether the friend is across the street or across the country. Further work marrying proximity, pricing, and alternate networks (like Wi-Fi) might take some of the strain off mobile users' data connections.

Microtelcos

Another approach to increase access is to relax national regulatory plans to allow for "microtelcos" or other local operators to be exempt from national regulatory pricing guidelines, or create them with specific mandates to provide access to rural or underserved areas, or both.[87] Trials in Papua New Guinea[88] indicate the potential for voice and SMS telephony to extend locally to very small communities, using inexpensive and small base stations. These microtelco models could be extended to cellular data connections with 2.5G, 3G, or even 4G connections.[89] This is a case where what is technically feasible with existing hardware and software may not be permitted due to regulatory structures and spectrum licensing agreements in place at the national level.

Alternatively, ISPs or communities can deploy WiMAX (the catchy name for the Institute of Electrical and Electronics Engineers's [IEEE] standard 802.16 instead of Wi-Fi's IEEE 802.11) stations to cover villages and neighborhoods with a single station. WiMAX was popular before the arrival of 4G LTE, but still may be appropriate for certain scenarios in underserved areas.

Look, Up in the Sky!

Although certainly "wireless," traditional satellite technology for Internet data transfer has suffered from high costs and long latency (due to long trips to the satellite) that have left it a viable option only for specialized users in underserved areas, with deep pockets and extreme connectivity needs.[90] Yet by 2014 there were new lower-orbit, low-latency satellite networks under exploration and/or early development including by Google,[91] Outernet,[92] SpaceX,[93] and O3B Networks,[94] all of which aim to make satellite-based data services and/or Internet more affordable.

Meanwhile, two high-profile global firms were reimagining connectivity for rural and underserved communities, served from on high, though not quite from space. In 2013, Google announced it was working on Project Loon. *Wired* magazine explains, "Project Loon balloons would circle the globe in rings, connecting wirelessly to the Internet via a handful of ground stations, and pass signals to one another in a kind of daisy chain. Each would act as a wireless station for an area about twenty-five miles in diameter below it, using a variant of Wi-Fi to provide broadband to anyone with a Google-issued antenna."[95]

In 2014, Facebook acquired a solar-powered drone maker, Ascenta, and announced its plans (in concert with its connectivity consortium, Internet. org) to explore new ways of providing Internet connectivity to rural and undeserved areas. As the *New York Times* explains, "The company envisions drones that could stay aloft for months, even years, at a time at an altitude of more than twelve miles from the surface of the earth—far above other planes and the ever-changing weather."[96]

At the time of writing, details on how either of these plans would work with existing handsets and devices are still scarce, to say nothing of details about pricing. However, this early stage and, on balance, the sentiments reflected in these ambitious projects are both laudable and worthy of additional scrutiny. At minimum, they are indicative of the depth of the frustration that some large Internet services companies (with significant revenues coming from advertising) have with the current status quo vis-à-vis access, and their enthusiasm hints at new paths beyond the cellular paradigm.

Conclusion

This chapter captured the great strides and continued challenges in providing Internet access to those still excluded—and there are many. No one approach will solve the access challenge for everyone. Rather, it will be a collaborative effort including multinationals, national network operators, ISPs, startups, regulators, social enterprises, and nongovernmental organizations (NGOs). Under the right market and regulatory conditions, advances in low-cost hardware, bandwidth-aware software, more extensive networks, affordable tariffs, and alternative networks (from libraries and Wi-Fi, to balloons and flying drones) all might play a role.

These impressive past-and-future advances in coverage are part of the reason I choose to call the book *After Access*. Coverage is not yet universal—nor are devices, or the skills or motivations to use them—but never before have the infrastructural barriers to Internet access been so low, for so many. There will be lots to discuss in the chapters ahead about what comes after access, and about the persistent differences between theoretical access and "effective" Internet use, but it is fair to say that access is not quite the insurmountable challenge it used to be, back in the days of fixed-line telecommunications. The glass 85 percent full perspective suggests that for many in the Global South, an affordable Internet connection is now within arm's reach. The glass 15 percent empty perspective reminds us that, given the technical and economic realities of mobile networks, those seeking to further expand and deepen access are still fighting geography, physics, and economics as they consider if and how to bring affordable Internet services to resource-constrained communities.

3   Gradations of Mobile Internet Experiences

What happens after access? Not everyone under a mobile signal currently uses a mobile device;[1] not everyone who does use a device actually uses the Internet;[2] and not everyone who uses the Internet from a mobile device does so in the same way. This chapter will further explore the term "mobile Internet" by asking two questions: how mobile? and which Internet? The answers will introduce considerable gradations and varieties of mobile Internet, stretching and complicating the idea of access beyond what many readers might have considered previously. To conclude, the chapter will suggest that it is better to frame the recent boom as a shift to a *more mobile* Internet, rather than to try to discuss mobile Internet in isolation from its fixed antecedent.

A Google Scholar search will yield over three thousand articles with "mobile Internet" in the title alone, over 700 of which specify "*the* mobile Internet."[3] It is doubtful, however, that everyone has been writing about the same thing. "Mobile" modifies "Internet" in a variety of ways,[4] and a basic problem is one of focus: does one stress the visible devices, the invisible wireless networks that connect them to the Internet, or the use cases that put these connections in motion? The technologies we call "mobile" sit at the intersection of an entanglement of devices, software, networks, and contexts. I shall consider mobile first, and follow with Internet.

How Mobile?

Let us run quickly through the entanglement of devices, software, networks, and contexts to illustrate how hard it is to pin down what we really mean when we talk about a mobile Internet experience.

For example, consider the devices. Cellular telephones were by no means the first "mobile" communication technology—Paul Levinson suggests that honor probably belongs to (stone) tablets, à la Moses.[5] One of the first "portable" computers, the Osborne, went on the market in 1981—it weighed 24 pounds.[6] Contemporary PDAs, cameras, handheld game consoles, and so on are all portable communication technologies, but seem not to be called "mobiles." Laptops, particularly lightweight ones, are inherently mobile (portable) devices, even if they have more in common with desktop PCs than with cellular telephones. Tablets reside halfway between mobiles and laptops; "phablets" (smartphones with big screens) halfway between mobiles and tablets. Looking ahead, health monitors, smartwatches, eyeglass displays, and Internet-connected automobiles are already stretching the limits of what we might call mobile devices. Yet in the meantime, across much of the English-speaking world, billions of handsets using the cellular GSM and "code division multiple access" (CDMA) standards for wireless communication have been granted the simple moniker of "the mobile"—the current electronic facilitator of human mobility sine qua non. For now, not all portable digital devices are considered "mobiles."

Similarly, a device's software is not a clear marker of whether a device is a "mobile" or not. Granted, smartphones are frequently powered by one of several common operating systems, like iOS (Apple) and Android (Google and many variants). Although these operating systems may be optimized for phones and tablets, they pop up in other devices with distinctly non-"mobile" form factors, such as on "smart" televisions (Samsung running Android), set-top boxes (Apple TV running iOS), and even a few laptop devices with keyboards and large screens.[7] Conversely, very "portable" laptops and tablets running PC operating systems (particularly Microsoft Windows 8.1) are very common, and the lines will blur further as devices begin to run Microsoft Windows 10, which will allow devices from handsets to PCs to share a common core of Windows functionality, and to run the same apps.[8] Proposals are also afoot to run a full desktop version of Linux on a mobile device.[9]

Network connections are not a great standalone marker of mobility, either. Of course, the cellular mobile providers provide widespread coverage to billions of devices using networks of towers supporting GSM, CDMA, HSDPA, HSPA+, and LTE network protocols. Yet many mobile devices also can connect via Wi-Fi to local fixed connections to the Internet. Many tablets do not even have a cellular network connection, and the devices themselves can still be useful even when a wireless network connection is unavailable. Conversely, many people use USB dongles, tethered mobile phones, or 3G modems to connect fixed PCs and home local area networks (LANs) to the Internet via the cellular network.[10]

Finally, and perhaps most problematically, there is the matter of context and intent—what users actually do with digital technologies regardless of their intended use case. Some theoretical literature treats mobility as a state of being, a condition in which people are in motion, relative to each other, their social contexts, and the places involving them.[11] In this sense, technologies are not mobile but rather can directly enable human mobility, or can complement human mobility by being portable. However, pinning any definition of a "mobile Internet" too closely to actual human mobility can be problematic because, regardless of location or portability, people everywhere frequently use portable devices to connect to the Internet, even when sedentary and particularly while at home.[12] For example, respondents in one study made 77 percent of their Google searches from mobile devices while at home or in their workplace.[13] This is not to say that the interplay between technology and human mobility does not matter tremendously. A broad literature now describes how even the basic mobile phone disrupts physical barriers, complicates space and time,[14] and encourages individuality, coordination, cohesion, and intimacy.[15] It connects people so seamlessly that Rich Ling argues we now take each other's reachability "for granted,"[16] and I will build on these ideas in chapter 5. Yet particularly in the case of the developing world, an exclusive focus on Internet use *on the go* would obscure other important factors surrounding mobile Internet experiences.

Which Internet?

The tour through devices, operating systems, networks, and intent indicates that there is no bright line between a mobile scenario and a fixed one. There are instead numerous gradations and variations, in which the traditional mobile telephone, connecting over a cellular network, is simply the most common, most recognizable mode of mobile communication. This next section turns to the idea of "Internet," suggesting that there are as many variants and gradations to the "Internet" part of the term "mobile Internet" as to the "mobile" part.

Smart, Feature, Basic

Non-voice (data) services on cellular networks have been available since the invention of short message services (SMS) in the late 1980s.[17] The wireless access protocol (WAP) found its way onto some handsets and networks beginning in 1999,[18] while an alternative standard for packet-switched data went mainstream in Japan with the introduction of i-mode.[19]

The period from 2003 to 2007 saw a rapid release of more powerful data-enabled handsets with increasingly faster network connections allowing users to access more of the Internet. Devices from Nokia, Palm, RIM (Black-Berry), and others supported early mobile operating systems including Symbian, Palm OS, and Windows Mobile, each allowing users to install and run software (later called apps) to expand the capabilities of the devices and to take advantage of the devices' data connectivity. Apple introduced the iPhone in 2007, and devices running Android, backed by Google, began appearing in 2008. As the browsing and network performance on these devices, which came to be known as "smartphones," improved, their popularity exploded.[20]

Today, handset manufacturers continue to experiment with different form factors and combinations of technical capabilities, all against a steady progression of technical possibilities in terms of raw processing power, battery life, miniaturization, display quality, imaging, sensors, software capabilities, and network options. The industry often uses a brusque hierarchy to divide mobile handsets into "dumb" (basic), "feature" (somewhat more powerful), and "smart" (most powerful) phones; figure 3.1 shows a 2014 version of each, side-by-side.

In less than a decade, smartphones have become one of the icons of our age, a symbol of "media convergence par excellence"[21] and the ideal postmodern device.[22] Smartphones are increasingly powerful handheld computers, running mobile-optimized operating systems including Android (Google), iOS (Apple), Windows Phone (Microsoft), Blackberry (RIM), bada and Tizen (Samsung), and Firefox. The major smartphone operating systems receive frequent updates, so not all devices necessarily run the same version, with the same features and/or compatibilities. There are also differences in the availability of apps. Across platforms, in 2014, Apple's iOS and

Figure 3.1
An example of a basic, feature, and smart phone.

Google's Android boasted a larger array of third-party apps (over a million in each case) than were available for other smartphone platforms.[23] Regardless, as a class, smartphone OSs are more powerful and easier to navigate, and host more plentiful third-party applications than any other mobile Internet devices.

Another group of handsets has become known as *feature phones*, which offer a limited set of features relative to smartphones, but at a more affordable price point, generally $40 to $80. The line between a smartphone and a feature phone is blurry.[24] Some feature phones have numeric keypads instead of touchscreens or QWERTY keyboards. Some feature phones support fast 3G or 4G connections, but others support only the slower 2.5G EDGE protocol. All can store and play media files like photos and short videos. What is critical for this discussion is that while they do not run a smartphone operating system like Android or iOS, virtually every feature phone can still access the Internet via a data channel, either through a browser or via downloadable apps. Using the data connection, users can do web searches, access stripped-down, mobile-optimized .mobi and WAP sites, and download games and programs from app store-like sites like

GetJar.com or Mobiles24.com. In 2013 Facebook claimed to have over a hundred million users accessing its services via more than three thousand models of feature phones.[25] Given their low cost, feature phones remain compelling and high-volume sellers in the developing world.

There are also still many *basic phones* in use. Even without a browser or a data connection, basic phones let their users make and receive phone calls, send and receive text messages, and exchange other data with their wireless carrier via the USSD protocol (an alternative to SMS). Perhaps the most famous basic phone in the world is the Nokia 1100, now discontinued but probably the best-selling mobile device of all time.[26] It remains the archetypal phone in the minds of many development practitioners, and may be the device one might be most likely to find still working—flashlight and all—in the dusty hands of a farmer or presented for (re)sale on a blanket in a local market in a small town. Today, its bare-bones successors sell for $15 to $20.

State-of-the-art, $500 "flagship" smartphone handsets will not be the center of the story in the developing world. Granted, $500 buys more power and functionality each year, but $500 is greater than the total monthly income for perhaps 70 percent of the world's population. By contrast, less than 20 percent of the world earns less than $50 per month.[27] Put in those admittedly coarse terms—I am not sure a full month's income is the right affordability threshold—a $50 handset is theoretically accessible to billions more people than a $500 handset.

Instead, the key developments to track in handsets for the Global South are at the $20, $50, and $100 marks. In 2012–2013, sales of new smartphones exceeded sales of new feature phones and basic phones combined.[28] These brisk smartphone sales are transforming the installed base of phones in operation. Ericsson projects that by 2018, 4.5 million of 9.1 billion total mobile subscriptions in use will belong to smartphones.[29]

Prices are falling. Some estimates suggest average smartphone costs dropped from $430 to $335 between 2008 and 2013.[30] In the critical $50 to $100 range, there is a contest for the attention of developing-world users between highly capable feature phones and increasingly affordable and powerful entry-level smartphones.[31] For example, in mid-2011, Kenya's Safaricom offered a low-end smartphone model, the IDEOS by Huawei, for $80, quickly selling hundreds of thousands of units;[32] by 2013, that device was available for $50.[33] Also in 2013, Nokia sold Asha feature phones for

$99.[34] By 2014, Mozilla had entered the mobile OS fray with devices like the Intex Cloud FX, a $35 smartphone for the Indian market. One blogger called it "2007's technology, today,"[35] but that $35 of functionality would have cost ten times that back in 2007. Google's Android is often found on lower-cost devices—many already under $100—that have helped Android approach one billion devices in use.[36] For example, in 2014, the Karbonn Smart A50S, possibly India's cheapest Android phone at the time, sold for $46.[37] Despite the compromises on hardware and software, a $50 smartphone is an impressive development, indeed.

The proportions of devices that are "smart," "feature," and "basic" will shift, and the capabilities of the devices themselves will be improving, but in the developing world in particular, gradations will remain. The optimistic take on this is that by 2020, perhaps half the handsets in the developing world will be smartphones (albeit probably still in the hands of the younger and the relatively prosperous).[38] The more cautious interpretation of the same trend is that purchasing power matters, so even the wave of $50 smartphones will remain out of reach for many. Smart is gaining, but among the billions with under $50 to spend on a handset, basic and feature phones are hanging on.

It is worth mentioning two other form factors at this time—tablets and phablets. Tablets bridge the line between smartphones and laptop computers. They sell well in the United States and other markets in the Global North, and are beginning to make inroads in developing countries. Yet Ericsson projects that the total installed base for tablets will be less than one tenth of that of mobile telephones through at least 2020,[39] a trend also visible in the small proportion of visits to websites from tablets captured in figure 2.2. Phablets, on the other hand, seem very much the rage, and it is no longer shocking to see people walking down the street with a device the size of a dessert plate held to the side of their head.[40]

Network Gradations

Another source of variability in mobile Internet experiences comes from the speed, capacity, and reliability of the network(s) users can access. A 4G-capable smartphone cannot consume much more cellular data than a feature phone if its user is in a rural area with only a 2G signal! In table 3.1, van Hooft offers a rundown of what users can accomplish at different levels of wireless access, contrasting them with fiber to the home (FTTH).[41] The

Table 3.1
Broadband applications using different technologies

	Fibre	2G		3G**		LTE***
	FTTH	GPRS	EDGE	HSDPA	HSPA+	
Max throughput*	100 Mbps	0.08 Mbps	0.23 Mbps	2 Mbps	56 Mbps	100 Mbps
Email	✓	✓	✓	✓	✓	✓
Basic Internet	✓	*	✓	✓	✓	✓
e-Govt	✓	*	✓	✓	✓	✓
Basic e-Health	✓	*	✓	✓	✓	✓
e-banking	✓	*	✓	✓	✓	✓
Music download	✓	*	*	✓	✓	✓
Video download	✓	*	*	✓	✓	✓
Tele-working	✓	*	*	*	✓	✓
Advanced e-Health	✓	*	*	*	✓	✓
Online gaming	✓	*	*	*	*	✓
High-definition IPTV	✓	*	*	*	*	✓
On-demand multichannel IPTV	✓	*	*	*	*	*

* Theoretical maximum downlink speed ** Downlink speed of typical HSDPA network deployed currently. HSPA+ theoretical maximum *** 20 MHz carrier. *Source:* Luke van Hooft, "Building Next Generation Broadband Networks in Emerging Markets," *Making Broadband Accessible for All—Policy Papers Series #12* (London: Vodafone Group, May 2011).

delineation of specific activity categories is perhaps a bit vague, but the general point is valid: 2G and 2.5G EDGE protocols handle text-based Internet interactions quite well, but for multimedia, particularly video, a 3G connection offers at least ten times the maximum throughput (speed), and is thus much more desirable.

Video is a particularly challenging scenario for MNOs and users alike.[42] Ericsson estimates that video content was 45 percent of total mobile traffic in 2014, projecting a rise to 55 percent by 2020.[43] Cisco estimates video consumed half of the total mobile bandwidth in 2013, and forecasts a climb to two-thirds by 2018.[44] YouTube alone comprises 40–60 percent of mobile video traffic in many markets.[45]

As is the case with handsets, estimating the exact numbers of unique 2G versus 3G *users* is challenging, partly because people can move between areas with different coverage levels several times a day. Yet the trends are evident. There is more 3G coverage to match lower-cost 3G handsets becoming available, and year-on-year growth in data usage (not users) exceeding 100 percent or more in many markets.[46] As I mentioned, Ericsson projects that by 2020, 90 percent of the world's population will have access to 3G broadband wireless connections, and 70 percent will have 4G connections,[47] though coverage in rural areas with sparse, low-income populations will lag urban areas.[48]

Indirect Access

One problem with table 3.1, or indeed with any discussion focused on higher-speed broadband connections, is that it obscures the ways in which even a basic phone can still be used to interact with data on Internet servers, via ubiquitous tools including SMS, USSD, and voice. In theory, these make any mobile handset, even a basic one with no data connection at all, a *potential* Internet connection via what I call "indirect" access.

The most important of these indirect channels may be SMS. SMS was built into the early GSM standard in the late 1980s as an early "data" service,[49] as a way for engineers to communicate with each other, but by 1997 European teenagers had discovered the channel, and the explosion of text messaging and, later, premium value-added services was underway.[50] The boom underway in 2015 of messaging apps such as Viber and WhatsApp probably mark the arrival of "peak SMS."[51]

Person-to-person SMS messages have their own rhythm and ritual—asynchronous, brief, and full of linguistic innovation[52]—yet the bulk of these messages never touch the Internet. Similarly, when SMS messages serve as the interface for premium services[53] like daily horoscopes or cricket scores, they allow the consumption of electronically stored data at a distance, but only within "walled gardens." These are powerful and flexible forms of mobile media, but are not the Internet.

More germane to this discussion are those cases when SMS directly interacts with Internet data via an SMS gateway.[54] For example, Google supports search, calendar, email, and blogging via SMS in various markets.[55] Wikipedia piloted an SMS interface with Airtel Africa in 2013.[56] Twitter's 140-character limit is testament to its interoperability with the humble text message. Innoz, an Indian company, has created an SMS platform to allow users to initiate a variety of Internet queries and functions (interacting with search engines, Wikipedia, etc.) via short codes and commands.[57] Finally, social enterprises like Frontline SMS[58] and the crowdsourcing tool Usha-hidi[59] make it easier for NGOs and development organizations to create linkages between SMS and Internet databases.

Sometimes, a "dumber" phone can use SMS to emulate a "smarter" one. For example, Mobile-XL offers its feature phone users, mostly in Africa, a downloadable J2ME app, or the app comes preinstalled on a SIM card. Launching Mobile-XL on a feature phone can give the user the impression he is using a slow, limited version of a smartphone, including versions of smartphone-style apps for Twitter, Facebook, and more. However, the app uses SMS behind the scenes to communicate with Mobile-XL servers.[60]

Similarly, the Unstructured Supplementary Service Data protocol (USSD) is a simple messaging protocol that works on almost any phone and is often free to use. For example, many MNOs use USSD to support airtime recharge and other functions. On my own network, pressing *141# tells me my mobile airtime and mobile data balances, via USSD. MNOs closely control USSD, and frequently require carrier partnerships before third parties can use the channel. However, like SMS, USSD can be stretched beyond its origi-nal intended use, to bring users in contact with Internet servers. Facebook rolled out a service using USSD on many Orange networks in Africa.[61] Meanwhile in 2014 Wikipedia worked on a USSD protocol[62] to allow MNOs to provide their basic phone users access to a simplified version of this quintessential Internet entity.

Finally, consider the case of interactive voice response (IVR). You have probably experienced an IVR when calling an airline or other large institu-tion. IVRs also support premium services like chat boards and movie list-ings. With the proper coding, these voice response (and/or recognition) systems can also facilitate reading from and writing to databases residing on Internet servers or on PCs connected to the Internet.[63] Such services may still be somewhat rare, but there is a great deal of excitement among practi-tioners in the development community about IVR and/or "spoken web"[64] channels. From agriculture[65] to livelihoods[66] to health,[67] voice creates new possibilities to extend some new modes of Internet interactivity to people who may have access to only the most basic of phones, and may lack the textual and technical literacies to manage traditional Internet- or even SMS-based text interactions.[68]

Many hope that clever two-way use of audio and voice can bypass and augment the screen entirely; hence some of IVR's appeal. For example, together with two colleagues and myself, Microsoft Research conducted a preliminary evaluation[69] of the "emergent practices" around CGNET Swara, a voice-based platform for citizen journalism[70] and community activism in the central Indian state of Chhattisgarh. In a community with no local-language news media, and very limited traditional Internet access, we found that CGNET Swara became an important channel connecting activists and citizens to one another and to institutional actors like local governments and police. Individuals were able to place a call to the CGNET Swara system and receive a call back, where they could then browse audio recordings on the system for local news, some songs, and a variety of stories contributed by other CGNET Swara participants. They could also leave a story of their own, in their own language. All the CGNET posts are stored on an Internet server located hundreds of miles away from the villages of Chhattisgarh. The editor selected some of the messages for transcription and translation, placing both the original audio file and the new text file on the CGNET website. Through a combination of technical and human transcription efforts, users' grievances, stories, and perspectives are online, not just in the original local-language audio form but also in translated, transcribed *text*, with the affordances of being indexed, searchable, and sharable just like other online content.

Regardless of current uptake, IVR, SMS, and USSD force us to stretch our ideas of what an Internet experience looks and feels like to users. In each case, anybody with a $15 phone and $.25 worth of airtime can conceivably interact with parts of the Internet. Although the popular statistics do not tend to count all of them as Internet users, all 3 billion-plus mobile handset owners already have in their hands an Internet terminal of a sort. It is just not the same Internet many of us would recognize, and it is certainly more constrained.

Indeed, we cannot really even limit the universe of people potentially affected by (and affecting) a more mobile Internet to those three billion or so who own a handset. I mentioned the prevalence of device sharing, both in families and via phone shops and public kiosks, in chapter 2. In addition, there is the case of *infomediaries* and *mediated use*, where one person uses a device on behalf of another.[71] If a teenager looks up a bus schedule on her phone on behalf of her mother who does not use the Internet, the teenager is acting as an infomediary, mediating the use of the Internet for her mother.

One example from a program to reduce infant mortality in Bihar, India, can illustrate several forms of indirect and mediated Internet use. In that program, the BBC Media Action group decided to use IVR content, accessed via mobile phones, to give community health workers scripts and information to share with expecting mothers. To make the program even easier to access and follow, they also printed large, colorful cards to accompany the IVR content.[72] At its core, this is an "Internet" project, since the content resides in the form of recorded messages from "Doctor Anita," hosted on an IVR server powered by the mobile-health application MOTECH.[73] Yet on the ground, it hardly looks like an Internet project. Is the community health worker using the Internet when she enters the mobile short code and hears the recorded voice of Doctor Anita? What about the expecting mother sitting next to the worker, simply listening to the voice while looking at the printed cards? Does simply hearing Doctor Anita, who resides on the Internet, count as Internet use? In this case, I would argue that the community health worker is the infomediary, using an indirect channel (IVR) to access Internet content, and that the expecting mother is also an Internet user, even if she never touches a keyboard herself.

Fringe cases like these are important in how we think about Internet experiences, more generally. From this perspective, almost anyone who knows someone who has a phone *can* interact with the Internet (even though many do not), and the maximum potential effective reach of the Internet is already somewhere between the 40 percent of the world's population who own a device and the 85 percent who live under a signal.[74] That is a remarkable development: we are accustomed to thinking of the Internet as having two to three billion users. If we include IVR, SMS, and USSD, it may already have four billion *potential* users, and, by adding mediated and indirect use, perhaps a billion more.

There Is No Mobile Internet . . .

It should be clear to readers by now that I do not think there is any distinct thing we can reliably label the "mobile Internet." Given the tangle of hardware, software, networks, and use cases I have described, a single useful, universal, and future-proof definition will probably remain elusive. Rather, the mobile Internet is, as mobile media scholar Gerard Goggin notes, "at best, a portmanteau term."[75] He instead suggests we consider a plurality of "mobile internets."[76]

Indeed there already is a multifaceted literature detailing how "the Internet" is not a singular concept at all, but rather a host of different experiences altered by, among other things, differences in language availability,[77] and emergent local norms and practices.[78] The expansion of mobile technologies has added to this list of ways in which "the Internet" varies: there are, again to echo Goggin,[79] many Internet experiences, and many of them are mobile. No wonder it is so hard to count mobile Internet users (or uses).

Nor are all mobile connections to the Internet the same. The original ideal of the Internet was for it to be a platform to connect equal actors via channels that gave almost no priority to one type of traffic over the other. It was a network of equals. Granted, by the end of the 1990s, differences among access speeds and hardware were already a feature of discourse on the digital divide,[80] but later chapters will illustrate how these gradations may be exacerbated in the era of a more mobile Internet. Those users with powerful smartphones and living under a 3G signal may have Internet experiences that replicate, rival, and in some cases even exceed the experiences available via fixed lines and personal computers. Those who have feature phones or only a 2G or 2.5G signal may have more *compromised* Internet experiences. Some will have only *indirect* mobile Internet experiences— reading and writing to Internet databases via IVR, SMS, or other non-IP services. Still others may be "users" but vicariously, interacting with a *shared* Internet via the handsets of friends, family, or other intermediaries.

It is also worth noting that some people, particularly prepay users on feature phones and 2.5G networks, running self-contained apps or downloading content directly from MNOs, might not even know they are using the Internet while they are consuming data.[81] My colleagues and I found this to be the case in interviews with mobile users in Cape Town in 2009, when some reported using Google or the popular mobile Java-based Mxit chat application on their feature phones, but assured us they were not "on the Internet."[82] Pearce and her colleagues found a similar scenario in Armenia, where survey respondents would list themselves as Skype and social networking users, then later report they were not Internet users.[83]

Consider the noteworthy explosion of chat and instant messenger apps, like WhatsApp, and WeChat. They are easy to use, easy to set up (linked to a mobile phone number rather than an email address), are light on data consumption, and nearly ubiquitous in some communities. They use Internet protocols to exchange data, but they do not resemble or require a browser session at all. Is it any wonder that the "Internet" part of the experience is not very salient? They might have gotten their start as ways to avoid costly SMS text messaging, but they are now a class of experiences unto themselves. Some are juggernauts. In early 2015, WhatsApp had 700 million monthly users sending thirty billion message per day.[84] Analysts suggest WeChat is used by 40 percent of all people online in the Asia Pacific region, with 400 million users in China alone.[85] To return to the theme of this section, is someone who uses WhatsApp or Mxit or WeChat on a feature phone, but has never seen a web page or sent an email, accessing or using "the Internet"? I would say "yes," despite the considerable strain such use places on the term.

. . . Nor a Clear Digital Divide for It to Close

Yet just as there is not a single mobile Internet, there is no clear digital divide for it to close. The evocative notion of a "digital divide"[86] has fascinated the academic, policy, and popular discourse around ICTs for decades. However, readers may notice my reluctance to invoke the notion of a digital divide throughout the book. Long before the shift began toward a more mobile Internet, there were already critiques of the arbitrary, binary nature of the concept of a digital divide, with scholars demanding better focus on multiple, interconnected divides,[87] or on reframing the divide as a spectrum of access and affordances, skills and opportunities.[88]

I will revisit these critiques in my discussion of "effective use" in chapter 8. However, my reluctance to use the term "digital divide" is also a result of the gradations exercise carried through this chapter. This chapter playfully suggests that we might want to consider 85 percent of the world's population as potential Internet users, given their proximity to cell signals and handsets. Yet for potential access to translate into use (in any sense we would recognize) remains an extremely tall order; it is still an outlandish oversimplification to lump users and potential users together in that way, especially when many people living under a mobile broadband signal remain "non-users."[89] I would say the same thing about the 45 percent who own handsets. There is simply too much gradation to the idea of access, now that cellular signals blanket the majority of the world's population, for a binary metric of access to be very useful. It does not make sense to talk about a family that shares a $10 cell phone and a family with a blazing-fast fiber-to-the-home (FTTH) connection and dozens of connected devices as being on the same side of any binary divide. Thus, I suggest avoiding the temptation to anoint any mobile Internet as *the* technology that will finally close *the* digital divide—for to do so is to accept a framing that a singular divide exists that can be closed. It is better, instead, to move to a formulation that allows us to see and address the persistent differences in Internet experiences among those 85 percent with theoretical access.

None of this discussion is intended to take away from a celebration of the state of the art. Yet it will be the $50 phone, not the $500 phone, that does the most to make the Internet inclusive for most of the world; conversely, the limitations of the $50 phone, from small screens and bad batteries to stale software and slow, intermittent, and expensive connections,[90] will structure the experience of many users. Even as "use" spreads widely around the world, we need to account for a greater, and more stratified, heterogeneity of Internet experiences than ever before.

4   A Lens on Mobile Internet Use

Although chapter 3 illustrated how a simple definition of "mobile Internet" is elusive, in this chapter we can turn the multidimensionality of the term to good use. I will introduce six elements that have each played a role in promoting the boom in uptake and use of a more mobile Internet in the Global South. Next, I will combine the six elements in order to offer a flexible way to distinguish mobile Internet experiences from previous fixed archetypes (in lieu of a simple definition of mobile Internet). I will conclude by converting the list into a lens that identifies specific implications of the shift to a more mobile Internet for development and digital inclusion in the Global South. This is the "After Access Lens" hinted at in the title of the book. The lens will structure all the chapters that follow, allowing me to illustrate how the same elements driving the shift to a more mobile Internet in the Global South bring specific new *mobilities*, as well as significant new *constraints* on effective use.

Before I proceed to these steps, I would like to spend a few paragraphs on nomenclature, and on why I am careful to describe what comes out of this chapter as neither a model nor a theory. Based on a reading of the ICT4D and mobile communication literatures and on my own fieldwork and previous writing, I have chosen the six elements that I think collectively best describe the particular boom in mobile Internet use in the Global South. There is no sophisticated semantic or statistical analysis underlying the selection or sequencing of these elements, nor a subsequent testing of their validity. I make no claims that this same list is applicable to any other technologies, past, present, or future. Thus, the particulars of this case, captured in the lens, may be of limited generalizable predictive value of the sort we should expect of social theories and models of technologies.

Instead it is merely a lens, albeit one about one of the most pervasive and transformative technologies ever to come into the world. Given the complexity and the nuances embedded in the rapidly changing technical landscape and popular term "mobile Internet," I think this categorization nevertheless offers a stronger and (somewhat) more durable perspective from which to consider the technology, compared to industry labels such as feature phones and smartphones, tablets and laptops, 2G and 3G, etcetera. I offer it in the hope that, as lenses can do, it will help narrow, focus, and sharpen the cacophony of signals and observable effects at cross-purposes that we confront when observing the shift to a more mobile Internet. Though the lens itself is not a generalizable theory or model, it echoes several such theories and perspectives in use among scholars interested in the interactions between technology and society.

Most central, perhaps, is Rogers' groundbreaking work in communication research on the *Diffusion of Innovations*, a frequent—albeit now perhaps somewhat settled—part of the ICT4D canon. The fifth edition on my shelf is from 2003, published forty-one years after the first. Its chapter 6, entitled "Attributes of Innovations and Their Rate of Adoption,"[1] summarizes a broad array of studies exploring how certain recurring attributes of technologies made them more or less likely to be adopted (used) by individuals, groups, and societies. These attributes: *Relative Advantage* (compared with alternatives), *Compatibility* (with existing practices and know-how), *Complexity* (or the lack thereof), *Trialability* (better a taste than a commitment), and *Observability* (better to be seen first, then done) can be used to predict adoption patterns of everything from cook stoves to safer sex practices (italics in the original). Says Rogers, "Cellular phones have an almost ideal set of perceived attributes, which is one reason for this innovation's very rapid rate of adoption."[2] Though I am interested in explaining uptake as part of the shift to a more mobile Internet, it did not make sense to use this framework as the primary organizing concept, since it downplays the specifics of any case in favor of a more general theory of adoption.

A wholly different perspective, also active in ICT4D, focuses on how technologies offer "affordances" for user action. Originally a term from psychology,[3] "affordance" was made popular in the design and human–computer interaction communities by Donald Norman[4]: "The term affordance refers to the perceived and actual properties of the thing, primarily those fundamental properties that determine just how that thing could possibly be used.... A chair affords ('is for') support and therefore, affords sitting. A chair can also be carried. Glass is for seeing through, and for breaking."[5]

Affordances were part of Norman's building blocks of a "psychology of everyday things" that also included constraints (what things cannot do, physically, logically, or due to cultural convention), and mappings (the ability of the user to link controls to actions). In the affordances approach, users synthesize perceived affordances, perceived constraints, and mappings between them to create "conceptual models" of how they might use an artifact or technology.[6] Researchers have applied the affordances approach extensively to explore and improve design practices, and to explain why users use technologies in the ways they do,[7] including mobile phones,[8] the Internet,[9] and social media.[10]

The affordances perspective has much to offer. However, as you will see from the list of elements in the After Access Lens, the explanation of the boom in mobile use in the Global South requires a lengthy consideration of costs (of installation, acquisition, and ongoing use) that I had difficulty mapping into the affordances perspective.

A third relevant perspective in ICT4D draws on the idea of materiality, from anthropology. Like affordances, materiality also stresses the interplay between properties of technologies/artifacts and the people that design, adopt, understand, co-opt, and use them.[11] In the introduction to *Invisible Users, Youth in the Internet Cafés of Urban Ghana*, Jenna Burrell describes materiality as stressing "the analysis of material forms and technologies of mediation in terms of their regular and specific qualities."[12] She suggests that the "principle of considering directly and at close range the technical and material as socially consequential is well worth keeping in mind."[13]

Media scholars from McLuhan onward have argued quite convincingly that different media forms influence, moderate, structure, and alter meaning itself—that it is important to be mindful of how the weather report "read" on a mobile phone is not the same weather report "read" on a newspaper or on a computer, or "heard" on the radio, even if the textual content of the reports is identical. Further, and just as important, the perspective asserts that a key to understanding the construction and meaning of these experiences is to situate these material elements of mediated experiences in particular places, cultures, contexts, and practices.[14] Yet, like the affordances perspective, the materiality perspective is not naturally amenable to explaining "adoption," particularly in universal terms. Readers will see some alignment between my perspective and materiality through the rest of the book. However, since these materiality approaches are deeply and necessarily situated in particular contexts, I do not foreground them.

Six Elements of Mobile (versus Fixed) Internet Technologies

So, I offer here six reasonably durable, reasonably pervasive elements of mobile Internet experiences. Taken separately, each partially distinguishes mobile Internet experiences from fixed Internet experiences, and each helps explain the massive uptake in mobile modes of Internet use around the world. Yet taken separately, each is incomplete, and no single element completely differentiates mobile from fixed. Only after I introduce all six can I begin to assemble them in different ways. Note that the first three are different elements of the cost of getting online via mobile technologies, versus fixed ones.

1. Inexpensive Devices

Mobile Internet experiences are possible on devices that often are less expensive than "fixed" devices—specifically PCs.[15] Device costs matter tremendously in determining the shape and scope of access. In many cases, people want and have mobile handsets anyway—to make phone calls, send a text, save photos, and more—so the incremental cost to a user of acquiring a device with a data connection is often zero. People sometimes start using the Internet long after purchase of the device.[16] A $75 entry-level smartphone, a $50 feature phone, and even a $20 basic phone can access the Internet (the latter by indirect means like IVR and SMS); all are less expensive than a new computer. From the diffusion of innovation perspective, the lower initial cost of devices offers a relative advantage of mobile Internet over fixed, and increases trialability.

2. Usage-Based Pricing

Compared to fixed connections, mobile Internet experiences occur more frequently under conditions of usage-based pricing. Prepay mobile voice tariffs launched in Mexico in 1992, then spread to Portugal, Italy, and beyond. Prepay uses unique codes, contained on printed scratch cards or generated at the time of a transaction by point-of-sale devices at retail counters, to allow customers to load "airtime" credits onto their SIM cards and phone numbers. As a user makes calls or sends text messages, her airtime balance drops until she reloads/recharges/tops-up her balance. When there is no airtime, it is impossible to place outgoing calls. By eliminating monthly bills, this approach unlocked the enormous pent-up demand for telecommunication services in the developing world.[17] Most consider it a win-win for MNOs and customers: prepay streamlines billing, making it much less costly for operators to serve customers. Customers no longer need a bank account, a steady address, or sterling credit to gain access to the network. Instead, they pay as they go, adding credits as small sachets of airtime, investing only as much in telecommunications services as they can afford that particular day or week.[18]

Usage-based pricing has made the jump from voice to data. Truly unlimited data plans are the exception outside of a few developed markets like Japan and South Korea; even in the U.S. most unlimited plans are being phased out. Instead, most users either have a monthly subscription with a data "cap"[19] or, especially in the developing world, pay by the bit in credits drawn directly from their voice airtime balance or in small bundles of data purchased in advance, like prepay voice.[20] Metered pricing on mobile networks is not a perfect solution for all involved—data can be more expensive on a per-bit basis than fixed channels,[21] and it can be frustrating to have to watch the meter for data consumption. I will return to these constraints at length in chapter 9. However, as was the case with voice, pay-as-you-go data fits the budgets, cash flows, and credit records (or lack thereof) of many resource-constrained users. In a diffusion sense, usage-based pricing of data is a relative advantage of mobile Internet over fixed, and increases trialability.

3. Wireless Connections

Mobile Internet connections usually depend on wireless network infrastructures—sometimes Wi-Fi but particularly cellular towers—to connect to the Internet. This wireless element has fueled mobile Internet adoption in two ways.

First again is the cost issue. It is important to stress the way wirelessness changed the economics of providing basic connectivity to households, businesses, and other locations, regardless of whether users are ever in motion or not. It is generally less expensive to install cellular towers in the middle of a populated area than it is to string or bury wires leading to every house, institution, and business in that same area.[22] This is particularly the case in the Global South where homes and organizations may not be able to afford expensive fixed lines, or companies may not decide to offer fixed alternatives at all, or both. Thus, wirelessness lowered the cost of initial connectivity to voice services, in most cases, to most users, and has helped make the initial costs of voice connectivity affordable for billions of people. The same is happening now with Internet access, as people who could not afford a landline broadband connection (dialup, ASDL, cable modem, or fiber) or costly satellite connections can now connect to the Internet via the cellular towers in their neighborhoods. In diffusion terms, the low initial costs of acquiring a wireless Internet connection (no modem necessary, if your device has a wireless radio) offers a relative advantage and increased trialability over fixed Internet.

Second, the anywhere-anytime nature of wireless connectivity, especially when coupled with a good battery, has been key in affording the shift from "place to place" to "person to person" connections, in increasing the ease with which individuals can contact each other regardless of location.[23] In diffusion terms, wireless Internet offers a massive relative advantage over fixed Internet on convenience and flexibility of connections.

4. Personal

Closely related to wirelessness is the fact that, compared to fixed connections, mobile Internet experiences are often personal and intimate. Users create strong bonds with their small devices, with screens and interfaces built for one user at a time, and often for one owner. Mobile telephones have been designed, sold, and bought from the beginning to be carried from place to place, going with us from home to work, from the subway to the park, living in our pockets and purses.[24]

By contrast, despite the "personal" moniker, PCs often belong to the room they are in (the living room, the office, the school, the shared-access telecenter) or to the families who bought them. While mobile devices can be shared, the norm is that they are not. As mobile devices gained data connectivity, applications and services taking advantage of personal ownership and intimate, private moments of use became common. I will explore the implications of wirelessness and personal ownership/control in much greater detail in the next part of the book.

5. Universal

The mobile telephone sold billions of units because, it seems, nearly everyone wants to connect and coordinate, to be in touch with loved ones, friends, and colleagues, and to feel secure in case of emergency.[25] Mobile phones interconnected with fixed-line public switched telephone networks (PSTNs), making them compatible (in the Rogers' diffusion sense of the word[26]) with existing practices for many with access to a landline, and substitutes for those without.[27] Even before the smartphone revolution, handsets had become impressive devices for entertainment and amusement, as well. From games and forwarded jokes to songs, cameras, and ringtones, handsets satisfied many desires that (albeit with lots of cultural and individual variation) could be argued were remarkably pervasive, almost essential to being human.[28] In 2007, Castells and his colleagues put it this way: "As elements of daily routine, wireless technologies, particularly the mobile phone, are perceived as essential instruments of contemporary life."[29]

The addition of Internet access has further expanded the degree to which mobile devices can appeal to near-universal human needs and desires. The technologies afford coordination[30] (making calls, sending texts, participating on social networks), as well as consumption of media in the form of images, text, sound, and video. From cats[31] and cricket scores, to pornography and pirated movies, there is something for everyone in mobile media.

One can see evidence of this universality in lists of the most popular apps in use. According to Apple, the top ten free apps on its smartphones in the United States in 2013 included four games, one media player (YouTube), one location utility (Google Maps), two photo-sharing social-networking sites (Snapchat and Instagram), one communicator (Skype), and of course Facebook (which is kind of all those things, again).[32] Similarly, in early 2013, mobile analytics company Flurry found that three categories—games, entertainment, and Facebook—together accounted for over 50 percent of total time spent on iOS devices in the United States.[33] In China, there is no Facebook, but Flurry observed roughly similar proportions there, with 76 percent of time on Android devices spent on games and entertainment.[34] We do not have comparable figures to these time studies available for app use by category in other markets in the developing world, but the trends should hold true. Chat, social networking, games, and media dominate mobile Internet use around the world, not just in the heavily researched U.S. market. For example, Souter found that in Kenya, Facebook was a significant driver of early mobile Internet adoption.[35]

6. Task-Supportive

The distinctive style of computing supported by apps deserves a special mention as one of the six core elements of mobile Internet experiences. Indeed, this is the only one of the six elements that does not have its roots in the technologies supporting mobile voice and SMS. (In other words, while the first five elements apply to any mobile phone, and were as applicable in 2004 as in 2014, this last element is specific to data-enabled feature phones and smartphones). Unlike on PCs, most mobile Internet experiences are carried out outside the browser, within apps (small programs) running on one of a handful of operating systems.[36] The same Flurry study mentioned previously suggests that users spend 80 percent of their time on Apple phones in apps, versus 20 percent on browsers.[37] The "outside the browser" behavior is due to a constellation of overlapping factors, including small-screen user interfaces that privilege one open application at a time; "locked" software and restricted terms of service vis-à-vis content ownership and software use; and concentrated value chains centered on real-time delivery of new apps through single electronic points of distribution (with accompanying editorial control).[38] In all, I call this element "task-supportive" design.

The implications of these concentrated ecosystems—open versus closed, browser versus app—are a contested domain in discussion of the Internet in general;[39] I will return to these issues in chapter 11. For now, we begin with the core observation that apps are *really* easy to use,[40] are *task supportive* on small screens, and are easy to purchase and update with "store" infrastructures behind them. From Angry Birds to Instagram, that ease seems also to be part of the appeal driving massive smartphone sales and mobile Internet use around the world. Paul Miller and Svitlana Matviyenko put it this way:

Every time you touch an app, you are basically just touching a metaphor, a conduit into an operating system linked to more metaphors layered on metaphor about the unfolding equations defining the data you see in front of you... As a conduit for digital information to invade every aspect of modern life, apps are scripts that link the way we search for small software solutions to things that we once took for granted, and to the larger issues facing a hyper-networked society on the precipice of total immersion in digital culture.[41]

In Rogers' diffusion model, complexity is an impediment to adoption. That is not a problem with mobile Internet use, compared to fixed. As I mentioned in chapter 3, apps are so in tune with devices' small screens, portability, and other affordances that people can interact with servers at a distance, reading and writing to Internet databases sometimes without knowing they are on "the Internet" at all.[42]

Fixed and Mobile Internet Antipodes

Of course, all six elements are not always present in every single device or use case. Imagine using a mobile 3G dongle to connect a desktop PC to the Internet, or using a tablet at home without an Internet connection, or using a mobile phone to project a video clip to a TV screen so that family members can all watch at once. These cases, and many others you might imagine from your own use of digital technologies, illustrate the lack of a bright line between fixed and mobile, and no single one of the elements I introduced above is sufficient to differentiate these cases as fixed or mobile. However, when aggregated, these six elements can portray mobile Internet use as a spectrum, with traditional fixed experiences at one pole, and a new purely mobile, "antipodal" archetype at the other. The more of these elements that are evident in a given experience (use case), the easier it is to describe the experience (use case) as a mobile one.

The differences between a tiny handset, a tablet, a notebook, a desktop, and everything in between are intuitive and, indeed, have evolved that way by design to serve different markets, needs, and use cases. However, these six elements represent design choices and infrastructure investments created and deployed by technology industries, and influenced by both policymakers and users themselves. Chapters 2 and 3 suggest that by numbers alone, the new mobile archetype is as common as the former fixed archetype, and as worthy of scrutiny.

I will grant that even the idea of a fixed Internet experience has shifted over time and varies enormously from user to user and from place to place. The earliest users in universities connected to each other via telnet and FTP; early users in homes in the United States and Europe may have used dialup modems to connect to bulletin boards, and later to services like Com-puServe and AOL.[43] However, for a long stretch, from roughly 1995 through 2010, the bulk of everyday interaction between people and the data and services on the Internet *was* via World Wide Web browsers on personal computer screens, whether in desktops or laptops. That Internet experi-ence, in which the Internet and World Wide Web were used rather inaccu-rately in the popular discourse as synonyms, came to define (or at least to dominate) the literature about the social and economic implications of Internet use. From home studies to school computer labs, from office cubi-cles to shared-access telecenters in the developing world, the idea of an Internet session having a start, middle, and end, as mediated through a browser on a personal computer became the unmodified "default." The literature is just now catching up to the fragmentation in Internet experiences.

I will also grant that the most significant argument against using a fixed versus mobile archetypal dichotomy to describe Internet use in the devel-oping world is the low-cost tablet.[44] Tablets borrow from the best of com-puting and telecommunications heritages, and from fixed and mobile form factors, with fewer and fewer compromises with each generation of release. Sub-$100 units are appearing, and although they are not yet a major factor in many emerging economies, several education departments have eagerly signed up for large-scale, low-cost tablet based initiatives like XO (One Lap-top Per Child) and India's Akash.[45] Yet even if affordable tablets sell in the hundreds of millions in the developing world, numerically they are unlikely to surpass the mobile handset as the first and most popular data-enabled device, particularly as phone screen sizes creep up over five inches into the realm of "phablets."[46] The 2014 Ericsson report, mentioned previously, projects less than a billion fixed broadband subscriptions in 2020, another 650 million devices classified as "mobile PCs, tablets, and mobile routers," and, at the same year, over *eight billion* broadband subscriptions for mobile telephones.[47] Nor are tablets exact substitutes for PCs; Gerpott and his colleagues found that tablet users in Germany consumed less data per month than laptop users, even when both groups were on similar flat-rate plans.[48]

Thus, we come full circle, albeit with several ambiguities acknowledged, and assert the importance of further detailed reflection on the experience of Internet use involving mobile technologies and infrastructures, particularly in the developing world. From here forward, unless otherwise specified, a *mobile* is a cellular telephone. I will describe the particular alignment of portable hardware, mobile software, connecting to the Internet over the cellular infrastructure across a variety of contexts and use cases. This is not because "mobiles" are inherently superior to other devices, but rather simply because *mobiles*, now more carefully defined, are so common. Until something dislodges the six (going on nine) billion cell phones from humanity's hands, it is worth focusing on the mobile/cellular handset as the emerging alternate pole on the Internet globe. The *data-enabled mobile is the antipode to the desktop PC's original pole.*

The device that has done the most to connect most of humanity to telecommunications services, in the historical equivalent of a blink of an eye, is not the telegraph, landline, satellite phone, or public telephone, but rather the affordable mobile. Similarly, the device that will do the most to connect that same segment of humanity to the Internet for the first time will not be a PC, a tablet, a phablet, a laptop, or a wearable device; it will be an inexpensive data-enabled mobile. These six elements, mostly inherited from basic mobile telephones and now deeply enmeshed in the shared infrastructure between voice telephony and data, will come to define the experience for hundreds of millions, perhaps billions, of new Internet users. Downloads and uploads, chat and photos, real-time crop prices and medical records, the profound and the mundane—all can be checked and updated on the go without a PC, by many who just a few years ago would have had no Internet access at all.

The After Access Lens

This chapter began by identifying a set of six elements that collectively account for the worldwide boom in access to and use of a more mobile Internet. We do not have to stop there. We can go beyond explaining the boom itself, and use the same set of elements to enumerate and explore the implications of expanded mobile Internet use for socioeconomic development and digital inclusion. As discussed, I call it the After Access Lens, set forth in figure 4.1, and its elaboration will drive the structure of the remainder of this book.

Part of the lens will tell the optimistic story of a technology that is providing more opportunities to people to interact with each other and

Six mobile elements

Social-structural implications		Inexpensive devices	Usage-based pricing	Wireless connections	Personal form factors	Universal needs	Task-supportive
With increased access...	Widespread adoption	✓	✓	✓	✓	✓	✓
Come new mobilities...	Place(less) interactions			✓	✓	✓	✓
	Place(full) interactions			✓	✓	✓	✓
...and specific constraints	Metered mindset		✓			✓	
	Restricted production scenarios	✓	✓		✓	✓	✓
	Circumscribed structural roles	✓	✓				✓

Figure 4.1
An "After Access Lens" on mobile Internet use.

share the world's knowledge. In particular, the mobile elements provide opportunities for individuals and communities to reconfigure their relationship to information and space in potentially powerful, productive ways. I will explore these new potentialities, which I will call place(less) and place(full) mediated communication and information behaviors, in part II of the book.

However, it is not a breathless story of universal, transformational success. Though the boom in mobile telephony has brought a step-change in the accessibility and affordability of Internet access and use, it has also brought increasing heterogeneity and significant gradations to Internet experiences. While a person with a $30 basic phone and a person with a smartphone *and* a state-of-the-art $1,000 desktop computer can each connect to the Internet, it may not be the *same* Internet.[49] The third part of the lens (and the book) will argue that, even "after access," these gradations in Internet devices and skills will both reflect and reinforce the broader challenges of supporting development and digital inclusion. My intention is not to temper the enthusiasm or discount the magnitude of the transformation underway, but rather to identify work still to be done.

The lens is complex enough to discourage any linear causal statements much beyond the surge in access itself. Any one of the six general elements associated with the growth in access is also responsible for some complicating factor, either transforming people's relationships to space and place through new mobilities, or introducing new constraints on certain types of effective use. Nevertheless, the lens is simple enough to be applicable across scenarios and contexts. Through promoting improvements to policy, applying social enterprise and civil society initiatives (including training), and continuing technical and business-model innovation, many parties have the power to influence, invent, and improve mobile Internet experiences in ways that favor digital inclusion. It is my hope that by the time we return to the full After Access Lens in chapter 12, its implications and utility for guiding theorization, policy, and practice will be apparent.

II   New Mobilities

5   Places and People, Switching and Serving

The previous chapters have explored the term "mobile Internet," uncovering what Herman, Hadlaw, and Swiss call a shifting, multifaceted "assemblage"[1] of technologies and activities that is pervasive, but ultimately hard to pin down. I suggested that while there is no distinct mobile Internet, we can still use "mobile" as a modifier—to describe the shift to "a more mobile Internet." The After Access Lens, described in chapter 4, introduces six common elements of mobile Internet experiences, each partially responsible for the boom in adoption and use in the Global South, and each with implications for socioeconomic development and inclusion.

Part II of the book will explore these implications in more detail. This chapter, in particular, will draw on existing research perspectives to situate a more mobile Internet at the intersection of three distinct discussions—about phones, about mobile phones, and about "the Internet" in general. The approach I have chosen is to isolate and then peel away these different, previous perspectives, exposing what is central to a relatively undertheorized "more mobile Internet" by the end of the chapter. Before I embark on this disaggregation, however, a few more caveats are in order.

First, as I hinted in the introduction, my approach linking properties of technologies to macrolevel social and economic changes is mildly technologically deterministic. It can be tempting to ascribe linear, predictable social change to the use of any given technology. Yet hard or reductionist determinism has long been suspect among many academic communities,[2] including ICT4D.[3] I do not want to suggest that mobile Internet use consistently—let alone universally—*causes* specific behaviors or outcomes in every case across all contexts. That said, the affordances and gradations of mobile Internet experiences I described in the previous chapters are present across multiple contexts because many elements of the technology are global by design and by market reality.

Let us return briefly to the example of a weather report discussed in chapter 4. Two fishermen in Benin and Peru might use the same handset, with hardware designed in Finland and manufactured in China, running an operating system created in the United States, linking to towers supported by technologies from Sweden, to launch a weather application designed in France to query data hosted on a server in the United Kingdom. Of course, at the water's edge, the fishermen's contexts, behaviors, and cultures are not the same—markets, weather, techniques, and a thousand other things will be distinct between the cases.[4] Yet the handsets and supporting technologies at the center of their digital repertoires do share many attributes, having been designed and deployed guided by assumptions that they would be used to support similar behaviors,[5] and do, *ceteris paribus* (all things being equal), structure the experiences of mediated communication in ways that are recognizable and important in both Benin and Peru. If each fisherman uses a handset to get a weather forecast, some elements of each mobile Internet experience will be similar.

Thus, while a comprehensive representation of theories and perspectives that capture the cyclical tensions and complexities of the interactions between technologies and users is well beyond the scope of this one book, it is important to balance any identification of similarities across contexts with cognizance of differences. I strongly encourage readers to consider assertions I will make in parts II and III of the book as *complements* to perspectives that stress, for example, cross-cultural differences,[6] appropriation and invention,[7] transgression and cultural reinvention,[8] circumstantial particularity,[9] structuration,[10] and social construction.[11] There are limits to what a tool-based, instrumental perspective such as the one I offer here can bring to bear on the phenomenon of mobile Internet use, so it is important to read this work in tandem with other, more contextually situated treatments.

Second, I should stress that I do not intend parts II and III of the book to break new ground on the theories of mobility or mobile interfaces *in general*. My task is rather to explore linkages between behaviors (Internet use, structured by elements and gradations of the mobile technologies at hand) and broader outcomes (particularly ICT4D outcomes in resource-constrained settings). My contribution is one of synthesis and integration rather than new theory or empirical research.

Finally, I must acknowledge that the chapters in parts II and III of the book are wider than they are deep, touching multiple subdomains within ICT4D. They do not offer a review of all of the dynamics of the use of telephones, mobile phones, or the Internet as they might apply to outcomes in development and digital inclusion. Nor is this the place to sketch or engage with the even broader debates on the relationships between the Internet and twenty-first-century global capitalism[12] or the information society.[13] My task is not to assess whether the Internet in general is a force for inequality or prosperity (or both), or even merely an amplifier of all human intent.[14] The focus, instead, is on the underexamined and often underdefined implications for ICT4D of the particular alignment of mobile *and* Internet, or more specifically, of *the Internet, when mobile.* To do so, the next section of the chapter introduces a matrix to distinguish technologies that switch from those that serve, and those that connect places from those that connect people.

Independent Yet Co-present Transitions

Societies are reflected in and shaped by their communication systems. Wellman[15] offers a way of thinking about communicative eras, which began with proximate, face-to-face communication, expanded via the telephone to encompass place-to-place communication at a distance, and now include person-to-person connections enabled by computer-mediated communication, including mobile phones and wireless computing. Similarly, the "Space of Flows" underpinning the network society described by Castells[16] is a product of the unique logic of communication and information exchange via interconnected computer networks. I draw on a few of these approaches to communicative eras, illustratively, in table 5.1 in order to highlight the ways in which mobile handsets simultaneously offer access to two great global networks-of-networks: the venerable public switched telephone network (PSTN), and the dynamic, multifaceted Internet. Users might not care where the PSTN ends and the Internet begins, but the distinction is worth trying to make in theory, practice, and policy, and to retain as a way of differentiating *distinct transformative technological waves manifesting in the same device.* Thus, table 5.1 takes the "telephone call" as

Table 5.1
Two departures from the telephone call, with illustrative themes

	Switching	Serving
Person-to-person	**"Mobile communication"** 1990+ Anyone can reach anyone else any time and any place	**"A more mobile Internet"** 2000+ Place(less) informational interactions Place(full) informational interactions
Place-to-place	**"The telephone call"** 1875+ Information moves instantaneously between any two connected places; great potential for coordination and communication	**"Networked computing"** 1980+ Users can manipulate computers (storing information, running programs) at a distance; links begetting links increase richness of the whole

the baseline condition from which new technologies for mediated communication can be compared.

The Telephone Call

Various authors including de sola Pool and Fischer[17] have identified the phone call as a quintessentially modern activity, affording millions of people instantaneous coordination, information exchange, and relationships at a distance to a greater degree than any preceding communication technology (books, letters, smoke signals, etc.). James Beniger links the "Control Revolution" even further back to origins in the telegraph,[18] but through the twentieth century it was the PSTN, rather than the telegraph, that became stronger and more global. Each additional phone added to the value of the network by being reachable by any other phone, tying homes, organizations, and countries together as never before. This reachability was an affordance not just of the visible device but also in the network(s) supporting this, and at the core of those networks was the switch, joining the user of any phone on the network to the user of just about any other.

As such, the telephone and the PSTN have been a great boon for socioeconomic development. In a comprehensive review written in 1994 (pre-mobile phone, pre-World Wide Web) for the World Bank, Saunders, Warford, and Wellenieus outline the benefits of telecommunications to development as including "market information for buying and selling," "transport efficiency," reduction of "isolation and emergency security," and "coordination of international activity."[19] It's quite a list, and illustrative of how much of what today is mistakenly and popularly attributed to "the mobile phone" or "the Internet" vis-à-vis socioeconomic development can be linked back to the real-time switching and connecting afforded by a technology in use for over a hundred years.

Read-Write Servers

Other functions do not have such a long history. The first dimension in this exercise contrasts *switching* with *serving*. By this, I mean the features and protocols that allow a digital device to access, download, upload, and change digital information stored remotely. Dan Gillmor called the aggregation of these protocols the "Read-Write Web,"[20] but neither the list of actions nor even Gillmor's phrase really do justice to the myriad roles servers can play—including acting as virtual switches to connect people in conversations as already described. Servers open a universe of possibility in the asynchronous, additive, interactive, networked, distributed manipulation and exchange of information. Heeks and Duncombe defined ICTs as the "electronic means of capturing, processing, storing, and disseminating information."[21] Perhaps in this sense, the server is the "I" to the switch's "C" in ICTs. Intimately tied to (but distinct from) electronic communication at a distance, serving facilitates *networked* computing, not just computing. Servers allow asynchronous or immediate interactions between people, but also between more than one person simultaneously, between humans and machines, between humans and "content" (text, audio, video, data), and even between machines and machines.

Of course, the greatest network of servers in the world is the Internet.[22] Although it might be possible to engage in extended technical,[23] social, and even mystical explorations of the essential elements underpinning all Internet experiences, this would be beyond the scope of this book. "The Internet" is an ever-changing tangle of servers, services, and protocols. Suffice it to say that the Internet is far more diverse than the PSTN, with social and development implications first forecast and later explained in aggregate by Bell,[24] Dizard,[25] Castells,[26] Wellman,[27] Benkler,[28] and hundreds more scholars. Digital databases, services, platforms, and structures contain information about the real world, social and physical. The *network logic* of a society mediated not just by switches connecting people, but also by servers, databases, and online services is pervasive and powerful indeed. However, to return to the rationale for drawing the contrasts in the matrix in table 5.1, the network society could have emerged even if no one had invented mobile telephony and the world was still using exclusively "fixed" modes like dialup, DSL (digital subscriber line), cable, and fiber to connect users to the Internet.[29]

Mobile Communication

Let us shift back to switches, from servers, and consider the rise of the basic mobile telephone in the 1990s and 2000s. From early on, mobile/cellular networks were designed to play a switching, connecting role. Thanks to behind-the-scenes interconnection agreements and a common numbering nomenclature, users of mobile telephones could connect to users of fixed telephones, and vice versa. In addition, thanks to more interconnection agreements, users on one mobile operator's system could generally call users on another. Thus, what could have become a set of competing networks—mobile and fixed—instead grew rapidly together.[30]

Yet mobiles offered something landlines could not: the ability to connect to another person regardless of one's location in physical space. Mobiles support person-to-person connections rather than place-to-place connections.[31] Books with titles like *Perpetual Contact*,[32] *Smart Mobs*,[33] and *Personal, Portable, Pedestrian*[34] are examples of a rich literature, now more than a decade old, exploring the social implications of wireless, mobile, personal, mediated communication. As a descriptor, I particularly like the comprehensive title of another one in the genre, by Rich Ling and Scott Campbell: *The Reconstruction of Space and Time: Mobile Communication Practices*.[35] Researchers and the general public sensed that humanity's relationships to space and time are influenced by the capabilities of mobile communication, even relative to the fixed telephone, which itself was quite disruptive.[36] The linear "impacts" of these changing relationships were not always clear, but the complications were apparent. Once communications went "on the move,"[37] many social processes required fresh (and ongoing) rounds of renegotiation. Availability had to be rebalanced with privacy,[38] individuality with collectivism,[39] public with private,[40] and home space with workplace.[41] Although the pressures of fractured attention and what Kenneth Gergen calls absent presence remain daunting,[42] Richard Ling suggests that societies have begun to take this level of reachability for granted.[43]

In an update to Castells' earlier discussions of "The Space of Flows and Timeless Time," Castells, Fernández-Ardèvol, Qiu, and Sey explain:

Mobile communication devices link social practices in multiple places. Even if the majority of calls are to people living in the same town, and often in happening a nearby place (Fortunati 2005b), the space of social interaction becomes redefined by creating a subset of communication between people who use their space to build a network of communication with other spaces. Because mobile communication relentlessly changes the location reference, the space of the interaction is defined entirely within the flows of communication. People are here and there, in multiple heres and theres, in a relentless combination of places. But places do not disappear. Thus, in the practice of rendezvousing, people walk or travel toward their destination, while deciding which destination it is going to be on the basis of the instant communication in which they are engaged. Thus, places do exist, including homes and workplaces, but they exist as points of convergence in communication networks created and re-created by people's purposes.[44]

However, it is important to stress that the dynamics Castells and his colleagues describe were observed long before mobiles had data connections, and still apply today as long as one's mobile "phone number" (a venerable convention) is visible to the PSTN. The complication of place, time, and space by mobile technologies initially had little to do with the Internet.

A More Mobile Internet

We have arrived at the last quadrant of the table, where, thanks to the massive spread of data-enabled mobile devices and networks capable of linking mobile devices to the Internet, the separate trends of read-write networks (enabled by the Internet) and the mobile society (enabled by mobile/cellular networks and devices) are converging and intermingling anew. Yet this framing restricts the range of the social and economic transformations we can attribute directly and exclusively to a more mobile Internet. If connectivity comes from fixed lines, mobility from cell phones, and enhanced "read-write" information storage, retrieval, dissemination, and processing from fixed Internet connections, then only services and experiences that demand both mobility *and* an Internet connection can be put squarely and exclusively in this fourth box.

Four of the six elements of mobile ICTs, detailed in chapter 4 and in the After Access Lens, are directly related to matters of "place." The wirelessness of the connections, their personal intimacy and portability, their universality, and their task-supportive designs each play a role in helping people renegotiate the relationships among *individuals, organizations,* and *places* by accessing, manipulating, and creating data that structures and represents these spatial and temporal relationships in new ways. There is a rich discussion of these renegotiations taking place in the mobile communication literature, including extensive treatments of mobile interface theory,[45] "mobile mediality,"[46] hybrid space,[47] and networked place.[48] Jason Farman offers an integration of these several threads, delineating between the *"removal of geographic fixity* and the simultaneous *site specificity of data"* as "two key features of our current interactions with" mobile media.[49] I will build on this delineation in chapters 6 and 7, delineating place(less)ness from place(full)ness to explore each in more detail.

For now, we can begin by situating these features in the fourth quadrant in table 5.1. *Certis paribus,* two potentialities are enhanced when person-to-person connectivity is coupled with read-write access to servers. First, mobile communication increases the capacity for place(less) addressability and informational activity of people in a social system, in which behaviors are not as dependent on the physical location of their actors; using mobile communication to access information and services residing on remote servers can accelerate and deepen this place(less)ness. Second, and rather paradoxically, mobile forms of Internet access can simultaneously increase the power, utility, centrality, and importance of location-based, place(full) informational interactions, in which place is important, and mediated by wireless communication with remote servers containing information about the places the devices and their users happen to be, have been, and will be going next. I suggest these two kinds of interactions are possible across all gradations of mobile Internet experiences.

I intentionally use "increases" rather than softer words like "enables" or "allows"; this is the crux of the mild technological determinism I referenced before. The paradoxical partners of place(less)ness and place(full)ness are amplified by mobile links to Internet servers and services, and each has specific influences on macrolevel factors of interest to ICT4D. Examples include location-based services, or participatory citizen journalism, or augmented reality on the move. I will cover these and others in the two chapters that follow.

How to Focus on the Fourth Quadrant

Flexibility, Production, and Play

Notice that no quadrant of the table deals with specific goals or behaviors. Like telephone calls,[50] mobile phone calls,[51] and Internet server queries,[52] mobile network computing capabilities are designed for—and used to support—virtually everything human beings wants to do, from the playful to the deadly serious. The enthusiastic discussions around mobile phones in economic and social development is understandable, albeit at times breathless. Yet while a more mobile Internet can be used in the service of socioeconomic development, it will not be used *exclusively* for socioeconomic development. Social uses, like games, media, and social networking, bring many users online via their mobile phones for the first time and currently dominate mobile Internet usage time on many smartphones. Leisure and self-expression are not a byproduct or a distraction, they are core to the elements of mobile devices, and must be part of the frame when we assess mobile Internet use.[53]

Indeed, for individuals (users), behavioral lines between leisure and work, between the social and the instrumental, may be quite blurry.[54] Emails from customers will intermingle with emails from cousins; a parent-teacher meeting announced on WhatsApp will be wedged between two pictures of friends (or between a picture and an advertisement); an app for scheduling maternal care nursing visits will sit on a home screen next to an icon for a game like Angry Birds. In addition, many of the devices will move seamlessly among workplaces, intimate domestic spaces, and every public space in between.

Finding the developmental and inclusion impacts amid the cacophony of daily life will challenge methods and theory, especially as data-enabled mobiles permeate daily life. Yet, "noninstrumentality"[55] is not a negative, or a problem. Indeed, it can even help people learn skills. As Tacchi argues, it is only really a challenge to those with determinist approaches who seek to foreground simple, linear "4D" frames and narratives on complex real-world contexts.[56]

Yet, due to the blurring of applications, user practices, roles, functions, and institutional actors, and due to the compelling affordances of mobile devices for consumption, expression, and coordination, it is difficult to place boundaries between uses "for development" and uses "in developing countries." When the social networking tool Mxit supports remote counseling for drug users in South Africa,[57] or when Facebook allows activists to organize, or when Twitter speeds disaster response, the lines between an M4D (mobiles for development) intervention and everyday appropriation of mobile and mobile Internet for "development-friendly" pursuits are not clear. Nor are the 4D implications clear when online services accessed via mobile phones are used for bullying, organized political violence and recruitment, government surveillance, money laundering, bribery, or hate speech, or when the same person who used the phone to check farm prices spends too much of his limited funds on airtime.[58]

These examples reflect a tension between bottom-up, chaotic appropriation and top-down intentional 4D services and interventions.[59] However these tensions are also fertile arenas for the exploration of subtle forms of personal, social, institutional, and economic transformation as the logic of a more mobile Internet (borrowing from the phone, the mobile, and the fixed Internet) goes worldwide.

Heterogeneous Installed Bases

Another thing to take from table 5.1 and this exercise is how a more mobile Internet is still "the Internet," with overlapping services, data, and servers at its core. Many, though not all, services residing on Internet servers are being configured to support user interactions across multiple devices.

Consider Ushahidi, the mapping and crowdsourced incident service developed as a way to deal with unrest in the wake of the Kenyan elections of 2007.[60] As a crowdsourcing system, Ushahidi makes it easy to gather incident reports from contributors, putting them on dynamic digital maps to help in disaster relief or other coordination activities. In Haiti, after the devastating earthquake in January 2010, volunteers used Ushahidi to help coordinate rescue and recovery efforts. Ushahidi supports data gathering via Twitter, email, and the web, but in the case of the Haiti earthquake, the SMS short code (4636) facilitated the bulk of the reports.[61] Given the strain on electricity and telephony, and the prominence of the "dumb phone" in Haiti at the time, the density of SMS messages is not surprising, but it is notable. Once the short code was active, every Haitian with a cell phone and knowledge of the short code could participate in an Internet mapping activity. Over subsequent months, responders received over 100,000 SMSs, and added selected messages to Ushahidi.[62] By design, and in response to conditions in the communities in which it worked, Ushahidi's functionality spans basic mobile, fixed Internet, and mobile Internet cases. To echo the theme of indirect access I raised in chapter 3, should we consider those individuals who submitted incidence reports to Ushahidi via SMS to be "Internet users"? I think we should, but in so doing, we reveal how new mobile interfaces, not just apps on smartphones but those on feature phones, and SMS, voice, and USSD, are challenging the archetypes of how an Internet experience looks and feels.

Similarly, in the health domain, services like Dimagi's CommCare[63] and the open source project Open Data Kit[64] have created powerful and flexible applications to streamline data collection and health field-worker support on feature phones, smartphones, and tablets. These systems work on mobile devices, but each have a web interface, as well.

Sanner, Roland, and Braa build on ideas from Hanseth, describing this kind of approach as the *cultivation of a heterogeneous installed base:* "very large and complex information systems, also referred to as information infrastructures, are never designed or built from scratch, but always evolve through extensions and improvement (cultivation) of what is already in place (the installed base)."[65]

Ushahidi and Dimagi serve heterogeneous device environments. So do Facebook, Wikipedia, Google, and a host of other services that allow users to read and write to their databases regardless of the device, fixed or mobile, smart, feature, or basic, that they might be carrying. Hundreds, perhaps thousands of other Internet services are investing and retooling for multi-device scenarios,[66] so that users can interact via their mobile handsets, not just through mobile browsers but also by apps, and in some cases also by SMS, USSD, or by voice interfaces. Responsive design, in which a website presents itself differently (and legibly) in different configurations depending on the device that accesses it, is a notable trend in this direction.[67] The "front end" user experiences and available features may vary among devices and platforms, but behind the scenes, hidden to users, data is intermingling in servers irrespective of the interface used to write to them.

Digital Switches Are Still Switches

I want to pause to reflect once more on switching, as a function, which occupies two cells on the table in this chapter. I do not wish to imply that switching is going away. People originally operated telephonic switches by hand, unplugging cables from one location on a board to another in order to initiate a connection between two users of the fixed-line network.[68] Later, analogue switches replaced human operators, and in turn were replaced by digital switches. Thus, the distinction I draw for this theoretical exercise is no longer specifically about physical architecture or technical design, but rather about broad, idealized function. Simply, the switch is the technological process in the middle of a network that connects two individuals in real-time, across distances.

Jeffrey Sachs has observed that the basic presence of mobile-enabled voice telephony helps break down the isolation that up until recently characterized daily life throughout much of the developing world, and specifically related to health, allows individuals and healthcare facilities to call ambulances (coordinate at a distance),[69] order new pharmacy stocks, or talk to a doctor.[70] In these cases, wireless infrastructures supporting low-cost devices helped expand basic conductivity beyond what was possible with fixed telephone lines. For example, one interesting startup, Switchboard, refines and repurposes the general mobile infrastructure for health. Its software creates a closed network of national health care workers, registered by USSD, powered by bulk SMS and free calls (subsidized by mobile operators) to get more out of thinly stretched networks.[71] Yet in the case of the call to the ambulance or even in the case of a virtual, multisite organizational phone network, the key value in these cases is still place-to-place connectivity. It just happens that the places in question were only connectable via wireless technologies.

Thus, it is critical to remember that the switch is not synonymous with the PSTN. Voice over IP (VOIP) services, particularly Skype, are increasingly popular on both personal computers and mobile devices. Some estimates suggest that by 2010, VOIP already accounted for one quarter of the world's total "global telephony minutes."[72] To users, Skype may behave like a switch, albeit a digital one, in that it offers video as well as audio connections between people in real time. A disruption like Facebook offering similarly switched, real-time voice connections (in addition to its served timeline and newsfeed) could move even more traffic away from switched mobile voice calls.[73] Industry analyst Russell Southwood notes the concern for emerging markets providers, particularly those in Africa, whose revenues are still coming from "voice minutes": "no one can tell you how long it's going to be, but at some point all services—voice in particular—will be data."[74] Meanwhile mobile messaging apps (WhatsApp, Viber, Kik, etc.) are eating into SMS messaging. Convergence between traditional telephony and data services is coming.[75]

This is not to say that voice will remain important for everyone, forever. Many people, particularly youth,[76] are replacing some mobile voice calls with other forms of mobile communication (email, instant messaging, social network status updates, photo sharing, etc.). Yet despite these trends, billions of handsets in the "installed base" and hundreds of millions more still rolling off assembly lines support connections to the PSTN, and hundreds of mobile operators across the developing world earn substantial proportions of their revenues by selling prepaid voice minutes. The legacy of connectivity to the PSTN remains appealing for many new users; a voice call is expensive (relative to a text message or a bit of data) but it is easy, ubiquitous, reliable, and has low barriers to use due to language or technical skill. It also may convey emotions and rich nuance of meaning in ways some people may struggle to do with text. Thus, even as the PSTN slowly fades and a greater share of communication is mediated by the Internet, there will be a need for a switching function. Whether requesting an ambulance in an emergency,[77] or wishing a sibling a happy birthday, there are some things for which an immediate, switched connection, whether voice only or a video chat, may be better than anything left on a server for later collection and action.

Conclusion

The initial spread of landlines, Internet connections, and mobile phones each came in different bursts, in different eras. Thus, the delineation between these technologies as visible artifacts seemed relatively clear, at least to casual observers. Early mobile handsets were like "phones" for staying connected on the go, and computers were becoming the primary means for interacting with the Internet; there were different archetypal devices, different archetypal networks, different research literatures, different policy environments, and so on.

This chapter illustrated how the data-enabled mobile device complicates this easy distinction. First-time mobile phone buyers walk out of stores with sub-$100 devices allowing simultaneous access to the great networks of both the twentieth and early twenty-first centuries (the PSTN and the Internet). One complexity around discussing a data-enabled mobile handset in the developing world is that it is actually an icon for both centuries. When it acts like a telephone (or telegraph), it supports the coordination and information transfer at a distance iconic of the modern age. The next moment, when it acts like a computer terminal, it brings all the complexities of the postmodern, information age.[78] When someone buys a mobile handset, she can blend use of several old and new technologies,[79] and can light up all four quadrants in the matrix.

The matrix exercise around table 5.1 helps isolate what is truly new about mobile-enabled data devices. *The "new" elements of a more mobile Internet are those that help people renegotiate the relationships among individuals, organizations, and places by accessing, manipulating, and creating data that links all of these levels in new ways and combinations.* In the next chapters, I will explore these renegotiations in more detail, as place(less)ness and place(full)ness.

6   Place(less)ness and ICT4D

The matrix presented in chapter 5 identified two new potentialities specific to mobile Internet connections. It is not the network server logic, per se, that is new (the fixed Internet does that), nor is it the place-complicating logic of personal connection (the basic mobile does that): what has changed are the times, places, contexts, and conditions under which a person (or machine) might engage with remote servers. Chapters 6 and 7 involve the intersections of place-to-place connectivity and network computing, and their implications for ICT4D and digital inclusion. This chapter tackles one of these intersections, place(less)ness, involving, in Farman's terms, the "removal of geographic fixity," and the loosening of the influence of space and place on informational behaviors. Chapter 7 will tackle its paradoxical twin, the place(full)ness simultaneously afforded by the increased "site specificity of data."[1]

Informational Interactions, Regardless of Place

To be online, one used to have to be *somewhere*—in a computer center, an office, a classroom, a living room. With mobile Internet connections, that requirement of being somewhere in particular has been eased. One can be almost anywhere. A networked form of place(less)ness is possible.

Take the hypothetical case of a commuter on a bus, armed with a data-enabled mobile device and an active cell signal. He might receive a WhatsApp message from his partner, read email from his job, view photos posted to Facebook by his cousin in a different city, or download a podcast or new song. All those are consumption activities. Just as important, he might also send a WhatsApp message, compose an email, tag himself in the photo from his cousin, or rate a song on the streaming music site. All of these acts actually write new content (or context) up to servers and services residing on the Internet. None of these digital acts, interacting with remote servers filled with accumulated structure and details about his social and informational life, have anything to do with the immediate context of the bus or the street on which it happens to be traveling.

The mobile literature has been wrestling with the decoupling of place, context, and communicative acts for a long time,[2] but the point I explore in this part of the book is that a much greater range of place(less) mediated actions are possible now that our commuter can reach not only other people (by a switch) but also information and services (via servers).

Portable handsets and wireless data connections have allowed more people to be freer to interact with digital information in modes *less* clearly determined by the social and physical limits of place, their institutional context, even their domestic living arrangements. Examples include music on headphones (from precursors in the playful, space-renegotiating use of the Walkman[3] to today's streaming services) to medical instruments that put a clinic's worth of sensors on a smartphone.[4] Yet the same device that monitors one's health may also play games, or contain an e-book and a camera. Users are more able not only to use a single service independent of location, but also to switch more readily between two or more digital activities, in what Ling and I called "interlacing."[5]

To clarify, place(less) is not placeless. Place always almost matters somehow.[6] For example, the commuter might experience a song differently while on a speeding bus compared to a bus stuck in traffic, or while on a bus that is nearly empty compared to one that is crowded. Also, these modes of place(less) interaction with data must be negotiated with and supported by the people shaping norms of use in particular contexts and places. It is possible that societies or communities would settle on norms of behavior that do not encourage these reconfigurations of time and space. If everybody was expected to keep mobile handsets plugged into power sockets on the wall, or if everybody simply turned them off while moving from one destination to another, there would be no place(less) addressability. Similarly, classroom rules about wearing earbuds at the table during lectures, or cautious conventions about displaying flashy, expensive digital devices on crowded city buses,[7] can reinforce contextual constraints on technology use. The technical potentiality for place(less) informational activity does not always win out against social norms.

Negotiations notwithstanding, innumerable design decisions and infrastructure configurations have created mobile devices that do not just enable but also *encourage* interlaced, place(less) communication and information behaviors. Recall the general properties of mobile communication technologies described in chapter 4; people are increasingly engaging in such activities partly because mobile Internet experiences, *as designed and deployed,* afford

- wireless connection to the Internet, even while in motion;
- intimate, personal, individualized control of a device;
- universally desirable coordination and consumption activities;
- near effortless support of discrete digital tasks.

Smartphone users need only spend a little time reflecting on the notifications on their devices to see the degree of place(less)ness encouraged by the handset: many will have a constant parade of social network updates, instant messages, emails, new pictures, and other alerts encouraging users to take their attention away from the place they are in, and into the phone. Or, more precisely, to take their attention through the device, through the wireless ether to an informational interaction with data and systems somewhere on a remote server. This is both a potentiality and a problem of our mobile society. The ringing phone and chirping SMS have been joined by a motley band of additional competitors for one's attention.[8] What Gergen called "absent presence"[9] is all too easy to achieve, stripping us of engagement with the places where we happen to be.

Place(less)ness, Development, and Inclusion

Though the bus commuter could be anyone in any city with a cell signal, the rest of this chapter will focus on scenarios in resource-constrained communities that are of interest to ICT4D practice, and for the broader goal of increasing digital inclusion. In other words, I will review some of the ways in which the new place(less)ness enabled by a more mobile Internet might be harnessed for socioeconomic development. This review is illustrative rather than comprehensive.

Diasporas

Rich, multichannel, inexpensive communication and coordination have already been a boon to the hundreds of millions of people involved in international and regional *migration and diasporas*. An array of studies about the use of the landline,[10] the mobile phone,[11] and the Internet[12] by migrants and dispersed families illustrate how all these technologies present remarkable opportunities for people to bridge distances and stay connected.

Yet further amplification and control of place(less) informational activities may be possible via a more mobile Internet. Connections (via switches) can be made wherever and whenever migrants have time and a few minutes to make them; they can be recorded and cached (via servers) for consumption across time zones. Identities can be managed and reaffirmed through more extensive contact with cultures and content back "home." Pleas for help (and stories to be heard) can be posted on social networks, relatively free from surveillance and control by others. Indeed, people can even better coordinate and exchange their wages and remittances via the mobile device[13] rather than at the bank (more on mobile money to follow).

I chose migrants in particular as an example since they are a group that might not have easy access to the fixed Internet as do those for whom they work. Accessing the Internet via the $50 handset they probably do own, and probably do control, is an encouraging development. Of course, more powerful ICTs aren't always entirely welcome, and place(less)ness cuts two ways. Like the landline and PC antecedents,[14] data-enabled handsets may become ties that bind, demanding more engagement with family left behind,[15] or can lead to clustering into digital groups bounded by shared culture rather than geographic proximity.[16] At this stage, we can say only that the place(less) affordances of the data-enabled phone will further alter, and perhaps on balance will improve, the management of family ties and finances from afar.

Gender and Agency

There are shifts in power and agency happening due to changing patterns of mobile device ownership, control, and use within households, as well. The relationship between gender and ICT use in the Global South is a complex topic, with numerous and often conflicting interactions among technologies, identities, opportunities, literacies, norms, and socioeconomic roles; a full exploration of these interdependencies would require volumes, not a few paragraphs.[17] Yet the most important development vis-à-vis women and mobile phones may simply be the increased opportunity for anyone, regardless of gender, to gain access to telecommunications (and by extension, the Internet), driven by each of the six elements of mobile ICTs described in chapter 4. Inexpensive devices, usage-based pricing, wireless connections above every home, personal/small/portable form factors, universal appeals, and task-supportive, easy experiences are encouraging men and women alike to purchase and use mobiles. Nevertheless, there is a persistent and observable gap in ICT use between men and women in resource-constrained settings,[18] and this gap includes mobiles. A large study by the GSMA suggests that in 2014, 48 percent of men in low- and middle-income regions owned a phone, but only 41 percent of women owned phones. Put differently, women in low- and middle-income countries were 14 percent less likely than men to own phones; in South Asia, they were 38 percent less likely.[19] When and indeed whether these gaps will close entirely, or by how much, remains unknown. However, it is probably safe to argue that in the medium to long term, further gains in gender-equitable Internet access will owe more to mobiles than to PCs.[20]

The improvements may actually likely mask continued complexities in the gendered use of mobiles and, increasingly, the mobile Internet. For example, many women at the base of the pyramid are not as likely as men to use text-based or Internet features on the phones they have.[21] In addition, portability affords sharing; in rural Uganda, Burrell found this sharing rife with difficulties and not always leading to extended communicative options for women.[22] Chib and Chen describe further tensions felt by community health workers in Indonesia, caught between the autonomy afforded by the phone and the gender roles requiring more subordination.[23] Indeed, some researchers describe how phones can be simultaneously "emancipatory and divisive,"[24] reconfiguring "double work"[25] or sometimes reinforcing rather than upending gender roles—or both.[26] Doron and Jeffrey suggest that as countless acts of transgression and self-expression facilitated by the phone have "accumulated, like grains of sand on a wind-swept beach, the dunes of social practice began to shift. The shape they would take was unpredictable, but worth watching and studying."[27]

It is against this background of rising (but still not equal) access and against the complexities of these unpredictable and contradictory patterns of use that any *particular* benefits associated with mobile Internet use by women would have to be discussed. As Pearce puts it, "Without access, the potential for empowerment is a moot point."[28] However, here is where the After Access Lens and the isolation of the specific potentialities of a more mobile Internet can come into play. I would suggest that in certain regions and communities where there are large discrepancies in gender roles and agency, the place(less) informational interactions enabled specifically by a more mobile Internet (relative to a phone, a mobile phone, or a fixed/PC connection) may create opportunities to reduce place- and resource-based male control of ICTs. Because mobile Internet experiences can decouple Internet use from place and context, more data-enabled phones in the hands of more women will open up new opportunities for users in ways that touch on other ICT4D narratives. Jo Tacchi describes one young Indian woman's use this way: "offline her social life is highly monitored. Online she is free."[29] From new opportunities to access education[30] to better tools for running a business,[31] to even simply communicating with whom she wants, when and where she wants, the place(less) informational interactions enabled by more mobile Internet devices in the hands of women, controlled by women, bought by women—especially young women[32]—will enable use with fewer barriers than devices in homes, schools, or workplaces, or even public areas.

Learning on the Move

Place(less)ness can also play a role in enabling different forms of learning. In this case, I do not want to focus on the extended access to content or pedagogy that *any* Internet connection allows. In other words, the use of a mobile device in a classroom, no matter how innovative the form factor and interface,[33] is usually not a place(less) activity. Rather, we can focus on the way the place(less) affordances of mobile Internet applications and services can benefit "learners on the move,"[34] even in contexts where traditional modes of learning and instruction are falling short. Some interventions might take the form of afterschool or optional programs delivered via mobile devices,[35] media delivery to mobile handsets,[36] quiet moments for teaching at home between adult and child, sharing content on a phone,[37] English literacy games authored by teachers for students to play on their mobiles at night,[38] university coursework repackaged as a "massive open online course" (MOOC),[39] lifelong adult learning in informal settings,[40] and mobile games.[41] All of these compose parts of what some perceive as an overall shift from e-learning to m-learning (m = mobile), and, to stress the counterfactual, would not have been possible if the Internet had stayed placebound and fixed.

That said, hard evidence for the effectiveness of any kind of impact of m-learning on educational outcomes in the Global South remains scarce, and certainly lags the general enthusiasm for the idea, in principle and pilot. To make matters more complicated, outcomes will vary according to the learner, her skills, the subject, the service, the content/pedagogy, and also the specific elements of the medium/technology that are the *ceteris paribus* focus of this book. The disaggregation exercise in chapter 5 illustrates how we need further evidence, and reviews of the evidence, that *mobile* Internet use rather than mobile use[42] or Internet use yields better educational outcomes before we can say definitively that place(less) access to servers and services enables learning. My suspicion is that the affordances *are* there, and that place(less)ness has a role to play in making m-learning successful, but whether these affordances are sufficient to overcome limiting factors in the institutional and contextual environments remains an open question, important to pursue.

Wider, More Efficient Markets

The shift to a more mobile Internet also has implications for livelihoods and productivity. Estimates suggest there are over five-hundred million farms in the world, the vast majority of which are smallholder properties in the developing world;[43] similarly, there are perhaps four-hundred million microenterprises (with five or fewer employees) in the developing world.[44] Billions of people rely on the success of a small enterprise or smallholder farm to survive.

There are growing literatures on the use of basic mobile phones by microenterprises[45] and small farms.[46] These literatures illustrate the simultaneity of the phone as a means for multiple practices outlined in chapter 4. On the one hand, farmers and microentrepreneurs are adopting and using mobiles as *telephones*—calling and texting suppliers, customers, partners, and occasionally experts or other resources in the service of their livelihoods. In these cases, the literature generally suggests that mobile phones, like landlines before them,[47] lower the cost of information search and coordination, increase the productivity of economic activity, and improve the efficiency of markets.[48] Some markets (like those for perishable crops[49] or those with a variety of buyers, sellers, and transport options) are more sensitive to the introduction of mediated communications than others are, but in general, the voice call is intuitive and requires little training. In other words, people around the world can now use the mobile phone *as a phone* for that quintessentially modern and industrial necessity—coordinating commerce and production at a distance.[50]

Meanwhile, an array of specialty services including price information broadcasts,[51] inventory and budgeting tools,[52] agricultural extension and training, and weather forecasts have been developed for mobile handsets.[53] Although no single one of these "serving" initiatives have achieved the scale to rival the switched voice call in terms of value to microenterprises and farmers, collectively, their variety and reach is impressive, and reflects the diversity of applications for which software and services can be developed.

Some of these services are simple, one-way message blasts. Others are interactive systems, delivered via SMS or USSD, which do not involve Internet. Increasingly, however, systems for agriculture and microenterprises are designed for heterogeneous installed bases[54] as described in chapter 4. Esoko's[55] suite of agricultural information systems is a good example of this multidevice, multimodal approach. To help its clients improve agricultural value chains, their cloud-hosted servers link overseas buyers and intermediaries (accessing Esoko via PCs and web pages) with farmers and small traders, who can read and write to its services using SMS or an Android app.

Another example of this approach to device heterogeneity is CellBazaar in Bangladesh.[56] Originally an MIT Media Lab project, CellBazaar started in 2006 as an SMS-based platform for matching buyers and sellers, but it has added wireless application protocol (WAP), web, and voice interfaces, and has grown to serve at least five million users.[57] Like eBay in the U.S., marketplaces similar to CellBazaar are available in developing countries around the world, although perhaps are better at serving urban sectors and users than rural populations. Similarly, Cheki, a used-car classifieds site in Kenya and Nigeria, has one million users, mostly browsing via an Android app.[58] Also in Kenya, iCow Soko (Marketplace), launched in 2012[59] and allows farmers to sell livestock, goat milk, and trees direct to Nairobi-based consumers.

Microentrepreneurs and farmers are also appropriating general social networks like Facebook for business purposes, advertising services,[60] or even coordinating on commodity prices.[61] These uses mix the personal and vocational in ways aligned with the informality and intimacy of their small enterprises.[62]

Thus, the more mobile Internet may change the participation of microentrepreneurs and smallholder farmers in local and global markets. First, there is the simple step-change in Internet access afforded by the mobile channel. People have been writing down (mediating) prices and markets for almost as long as there have been prices and markets. However, the Internet has enabled prices and transactions to be created, updated, tracked, and used more dynamically than ever before.[63] Distinguishing traditional marketplaces from digital marketspaces, Rayport and Sviokla describe how "when buyer-seller transactions occur in an information defined arena . . . information about the product or service can be separated from the product or service itself."[64] Thus there is a place(less)ness about all electronic marketplaces, but the mobile handset will be the device that allows more microentrepreneurs and farmers to participate, even if they never leave the farm or their shop desk.

Beyond affording widespread access to the Internet, is the "mobile" in mobile electronic commerce important in the Global South? So far, the place(less) elements of mobile Internet use are probably not as important to most microenterprises and farmers as are the handset's ability to provide basic connectivity to the PSTN and Internet. However, new models of allocating labor,[65] particularly microtasks and crowdsourcing—using technology to divide a task into tiny bits and distribute to freelancers—are particularly well suited to place(less) use cases. These are not just mobile-friendly job boards, although those, too, are finding success in many markets in the developing world.[66] Instead, microtasks, freelancing, and crowdsourcing models use technology to transform the traditional employer-employee relationship. The best known of the commercial crowdsourcing services is probably Amazon's Mechanical Turk. Indeed, many of Amazon's Turk freelancers reside in India.[67] However, the extension of a paid crowdsourcing model to mobile handsets is new. The startups like MobileWorks[68] and Jana[69] face the heterogeneous device environments, serving as a web-based front end to task owners and a mobile handset-based interface to task workers, to put more opportunities in front of more people. Both startups emerged out of computer science labs at U.S. universities with programs looking at innovation for the developing world.

Any growth in mobile-enabled microtasking in the developing world may present a double-edged sword. Microtasks are certainly appealing as a form of "interlaced" pickup work during hours of idle time, but more evaluation is necessary to determine whether they can reliably replace a steady job, particularly if they are framed in that context in the rhetoric of social enterprise and global good. One counter-argument may be that the ruthlessness of a twenty-four-hour global business cycle may keep wages for microtasks well below what people with similar skills in a placebound labor context could command. In addition, persistent critiques about outsourcing as exploitation (rather than opportunity)[70] focus on the instability of the microtask approach. Pending further careful research, the overall effect of microtasking on livelihood opportunities for resource-constrained people in the developing world remains unknown.

Mobile Money

To complement the preceding section on markets, it is worth briefly mentioning mobile money.[71] Mobile operators in the Philippines and Kenya[72] demonstrated early successes with mobile money. Mobile money allowed users to "cash in" currency at stores where mobile airtime was sold, "store" that currency in accounts accessible via mobile handsets, "transfer" that currency to other users at low cost, and if necessary, "cash out" that currency at any shop in the same network.

Mobile money is spreading beyond these successes to many more markets, as an alternative or complement to scarce, expensive bank accounts. The technology is not "mobile Internet" per se, which is why it gets only short coverage in this chapter, but it is an important part of the emerging mobile services infrastructure. Mobile money provides greater control over financial identities to users, allows them to accumulate transaction histories and credit scores, and the place(less), low-cost person-to-person transfers it supports mean that companies like iCow Soko in Kenya can count on most of their customers having a means to pay for goods and services. Similarly, in the world's diasporas, mobile money provides a vehicle for sharing funds across space and time that is cheaper and more intimate than alternatives like Western Union and the post office. Mobile money will become intertwined with the mobile Internet as credit cards and other payment modes have become intertwined with the consumer-facing fixed Internet. As banks offer mobile money services, and as interconnections among mobile money and other payment networks become solidified, the same dynamic visible in mobile Internet trends is apparent here; mobile interfaces become new ways to access existing networks (in this case, financial networks) in heterogeneous device environments, rather than walled gardens unto themselves.

Conclusion

This chapter has offered a set of examples from domains of interest to ICT4D and inclusion. In each case, I have taken care to isolate the potential social-structural implications specific to the new mobilities (in this case, place[less]ness) made possible by the shift to a more mobile Internet. In the next chapter, I will cover place(full)ness in a similar way.

For now, the takeaways from the chapter are twofold: First, a greater sense of how the case of Internet access and use, via mobile devices, offer users increased flexibility to renegotiate and occasionally transcend their connection to particular places. To the extent physical mobility has always been a luxury of the prosperous and the powerful, one may hope that on balance, the spread of these new place(less)ness potentialities will result in more good outcomes than bad ones.

In a decade or so, it may be possible to look back on this period and identify major shifts in the macro- and micro-geographies of economics and participation in the Global South, thanks to a relaxation in the place-boundedness of informational activities. Yet the second takeaway is that this assertion is merely that—an assertion—pending significant further testing and exploration by the research community.

7 Place(full)ness and ICT4D

Chapter 6 focused on ways in which mobile Internet use supports place(less) informational behaviors—ones occurring with very little reference to the physical location or context of the user. Yet the inverse is also true; several elements of mobile communication and media (particularly mobile Internet) simultaneously increase the frequency and power of what Wilken calls "communication functionally bound to location"[1] and "precise geographical locatability."[2] Farman's term is *"site specificity of data."*[3] When a user is in a place, and her network-connected mobile device knows she is in that place, there are new opportunities for informational and communicative acts that specifically reference and integrate place and user presence. Katz and Lai review these developments as "Mobile Locative Media."[4] Some of this was possible even with basic mobile phones, but multimedia capabilities (image[5] and audio/video[6]), sensors, Internet connections, and the rise of "location-based services" (LBS)[7] offer a step-change in the depth and range of digitally mediated, place(full) activities.[8]

Consider the litany of location-aware applications available on smartphones. Users can navigate on foot, by transit, bike, or a car using map applications. They can "check in" to find nearby friends on social networking services, or search for romantic partners on location-aware dating services. They can check traffic and calculate fastest routes through a crowded city, or order a taxi or a shared ride. They can save and exchange routes they have walked, even in the wilderness, using GPS. They can "geotag" a photo, video, or message uploaded to a cloud account or social network. They can search nearby for shops and services via search engines; or have promotions from shops nearby sent to them via SMS. They can track a child in a playground, or a package in transit. They can play games that demand interaction with "real-world" landscapes and players,[9] or access stored historical and social information about a place left by others visiting before them.[10] They can also leave behind digital traces of their own interaction with a place. In stark contrast to Gergen's "absent presence,"[11] Farman reminds us that, "a person staring into a screen can actually be someone who is deeply connected to the space he or she is moving through."[12]

Thus, if it can be mapped, it probably has been mapped. If it can be tagged, it probably has been tagged. A flood of information about the physical and social worlds moves over Internet servers and databases each moment of each day. Multitudes of moments are situated with three-dimensional physical coordinates, time stamped, and augmented with what Mark Graham calls "virtual dimensions"[13] (relative readings on social, biological, and emotional states), aggregating both personal and "(im)personal geosocial information [from strangers]"[14] into power-dependent, contested[15] digital representations of places and social contexts in motion and flux. Remote Internet servers and systems hold, sort, and compile the data, but increasingly, the mobile is the device through which that information both is gathered and acted upon.

Some location-based services are worldwide phenomena; others are region-specific. Users appropriate and reimagine those services that make the most sense for their context and needs. See, for example, Hjorth and Gu's description of users of Jiepang (similar to the check-in service called foursquare) in urban China,[16] or Walton's contrast of class and vernacular reflected in the geotagged images uploaded to Flickr and the mobile social network "the Grid" in Cape Town—the two networks portrayed two very different perspectives on the very stratified city of Cape Town.[17] Clearly, research into the social implications and use of location-based services and space-specific mobile use is too broad for full coverage here, and I recommend other resources.[18] Instead, the goal of this chapter is to link a few of these place(full) informational behaviors to the social and economic outcomes of interest to the ICT4D discourse.

Protest and Pressure

Foremost in mind for many readers might be the role that social networking sites and other Internet services, as accessed by mobile devices, has played in coordinating and enabling political protests around the world. Whether evolutionary, in the form of citizen journalism, or revolutionary, in the form of mediated protest events, mobiles have a role in structuring what Rheingold calls "augmented political action,"[19] Castells calls the dynamics of power and counterpower,[20] and Bennett and Segerberg call "connected action."[21]

Protest

Note that mobile-powered protests, popularized by Rheingold's idea of *Smart Mobs*,[22] crossed into the public/political eye around the EDSA 2 protests in the Philippines against President Estrada[23] in 2001—long before the widespread availability of data-enabled handsets, and before the rise of what we now call social media: Twitter, Facebook, WeChat, and so on. Those protests, and others in the same period, were coordinated using voice calls and text messages, and were reflective of the switching functions I discussed in chapter 5.[24] Conversely, not all mediated protest *requires* a mobile component. Even a decade after EDSA, it is possible to do an analysis of online protest that does not explicitly call out the role of mobile devices, as separate from the Internet or social media.[25]

That said, there is a great deal of visibility of—and scholarly interest[26] in—the use of Internet-enabled mobile devices in protests, including those in Kenya,[27] Iran,[28] Spain,[29] Turkey,[30] Ukraine,[31] around the Occupy movement,[32] and in the various countries involved with the "Arab Spring."[33] In these accounts, and, more importantly, in these protests, mobile devices link events "on the ground" to servers and services on the Internet, which both capture and represent those events for participants and nonparticipants, and indeed for the world.

As Monterde and Postil note: "If a few years ago it was still justifiable to separate mobile and internet technologies when studying 'smart mobs' such as People Power II . . . . with the advent of smartphones this is no longer possible."[34] These new linkages are occurring alongside and in addition to the bulk SMS-driven coordination functions observed by Rheingold at the time he wrote *Smart Mobs*. Calling a friend to join you at the protest is an act of coordination (switching); so too is forwarding an SMS to dozens of friends at once, if the SMS disappears into to the vapor of time, unsearchable and undiscoverable. Posting status updates to your Facebook page about how many people you have invited, or imploring others on twitter (with a hashtag) to invite their friends, writes your experiences to a server that will hold that information for some or all to see, or even for an algorithm to evaluate, across time and space. It is something different from a switched phone call.

By combining on-the-go access to networks that switch *and* serve, the smartphone is a protest tool par excellence.[35] Four of the core elements from the After Access Lens are on display: protestors can (a) connect wirelessly to the Internet, from (b) inexpensive and (c) personal devices, to use (d) powerful task-based functions—in particular, location-based services, social networking services, and multimedia tools, in order to capture, edit, upload, and download images, audio, and video.[36]

Similarly, the protest venue is one of the best examples of the potential of the more mobile Internet to enable place(full) informational interactions between people and data. In some cases, the protests take the name of the space they are in, "occupying" geographies like Tahir Square in Egypt in 2011, or Gezi Park in Turkey in 2013. Yet even when the point of convocation has no name, place(full) behaviors enabled by a more mobile Internet are evident. Earl explains how sharing via Twitter locational information about protest activities (and the authorities' whereabouts in relationship to these activities) reduces "information asymmetries" between police and protestors.[37] Here the salience of the server versus the switch helps illustrate a shift, and a new set of powers. Protest participants are using data-enabled mobile devices not only to assemble and coordinate the protest, but also to structure and contest its message and meaning by creating, propagating, broadcasting, negotiating, and consuming digital information about the protest, in patterns that are faster moving and more dynamic than anything possible with previous technologies.[38]

Pressure

Meanwhile, independent of punctuated, tumultuous public protests, the work of building inclusive, responsive civil societies and increasing participation continues every day. NGOs and activists around the world are exploring ways of using ICTs to expand "voices" from people, communities, and perspectives traditionally excluded from the public arena. This activity is described in many ways, including citizen journalism,[39] participatory content creation,[40] community and citizen media,[41] transparency promotion,[42] and digital inclusion.[43] Many of these emerging approaches to digital citizen engagement are not specifically mobile, but rather are multimodal, interacting with "heterogeneous user bases" (as I described in chapter 4) with an increasingly active mobile component. Ushahidi's mapping platform, used to track everything from election violence in Kenya to corruption in Macedonia,[44] remains a good example of this heterogeneity, though other systems are proliferating.[45]

That said, a variety of specialized applications take specific advantage of mobile platforms to expand mediated discussions, improve real-time accountability, monitor elections,[46] enable coordination between activists, and influence public policies. Freedom Phone[47] and IVR Junction[48] are platforms for interactive voice response, with an eye toward supporting citizen journalists. India's "Bribecaster,"[49] "I Paid a Bribe,"[50] and other anticorruption apps are examples of mobile-specific services designed to make the most of the affordances of a more mobile Internet. Mills and his colleagues call the citizen journalism approaches in this vein "MoJo."[51] Qiu, meanwhile, identifies the rise of "worker generated content" in China, via smartphones, after a long history of local protests coordinated primarily through voice and SMS.[52]

Most of these citizen-engagement use cases do not necessarily have to be place(full); indeed some probably rely on place(less)ness as well, allowing journalists and citizens to post, browse, interact, and exert pressure on whatever topic they like, from wherever they are. Yet place(full) citizen engagement is possible. Pilot programs and experiments from the ICT4D community are illustrative, including election monitoring in Nigeria,[53] reporting road repair conditions in Kenya,[54] monitoring water quality[55] and updating land records in South Africa,[56] and enforcing teacher attendance via location-authenticated interactive voice response in India.[57]

I offer these examples not as a blind celebration of instant transparency and efficiency. Anytime the institutions change the way data is managed or shared, there will be new contests over power, winners and losers.[58] Nevertheless, the experiments underway suggest that the leveraging of place(full) data generated by billions of mobile phones will continue. ICT4D practitioners and institutions may be able to participate in the transition to more inclusive, more transparent systems, thanks to the integration of mobile channels and the place(full) informational interactions they enable.

Place(full) Informational Interactions in Protest and Pressure Are Observable, Even if Their Consequences Are Not, Yet

Does all this amount to anything? The debate about the overall effectiveness and centrality of mobile and Internet social media in "augmented revolutions"[59] and citizen engagement[60] is ongoing (often with conflicting definitions of *ICTs* and *impact*),[61] and a review of that debate is beyond the scope of this book.

Nor is all the place(full) ascription and information exchange an unqualified good; there is a risk that any (or every) tweet, geotag, post, association, and SMS will also be read by authorities, either at the moment of the protests or for years afterward.[62] For example, the government of Thailand asserted in 2014 that it had the capability to monitor all thirty million Thai users of Line, the popular mobile social-networking app.[63] The risks to privacy and safety by protesting with a smartphone have eroded anonymity, a theme I will take up again in chapter 11. Also, not all activities around protests are unequivocally "good" activities. Data-enabled mobile devices may help other groups coordinate destructive actions like riots and inter-ethnic conflicts[64] or protest suppression just as easily as they will help "the good guys" (whoever they are). That said, if extreme groups like Boko Haram and the Taliban[65] blow up cell towers, perhaps that is another form of evidence that coverage does more good than harm.

My narrower point is this: to the extent that social media and the Internet *do* play a role in changing political regimes, and in allowing citizens' voices to be heard or their will to be expressed, the connected mobile device is increasingly central in that role. That is why, for example, there were mobile charging stations set up at the Gezi park protests in Turkey[66] and the Euromaydan protests in Ukraine.[67] Mobile devices are important in protests not just because they are Internet connections, or phones, but specifically because they allow access and use of a more mobile, place(full) Internet. The Internet could not play the role it has in protests in recent years without the addition of *mobile* access.

Volunteered Geographic Information

Complaints and feedback about government services are just one form of information that can be gathered via mobile channels and mapped on the broader web. Adding more links between the Internet and places in the physical world is also proving useful in gathering widespread and dynamic information about many natural phenomena, from environmental conditions and air quality[68] to wildlife spotting[69] and even disease outbreaks. Mobile devices have helped facilitate what Elwood, Goodchild, and Sui refer to as "volunteered geographic information."[70]

Place(full) Value Chains

Many markets can be improved by adding geographic information to transactions via place(full) interactions with the more mobile Internet. For example, applications running on smartphones can tell consumers at the time and moment of possible purchase about the source of a product, the wage practices of its company, its environmental impact, and more.[71]

Goods on the move are place(full), too, and information about transfers along the value chain from the maintenance of refrigeration temperatures to inventory management can help improve logistics and transport infrastructures in low-resource areas. This is as important for vaccines and medical supplies as it is for cut flowers or auto parts. Mobile-based Internet technologies from barcode scanners and UPC codes to sensors and manual data-entry forms will help improve all manner of transport logistics.[72] These include provisioning prepaid and solar electricity from utilities via mobile meters.[73]

One common scenario involves traders and business on the move; the basic mobile phone has already been a boon to many of them, allowing better coordination of activities and dynamic matchmaking between buyers and sellers.[74] The mobile Internet promises to amplify these trends. As read-write databases augment simple switches, a host of just-in-time business and sharing models already popular in the cities of the Global North will spread further. Transport companies like Uber are spreading into the cities of the developing world, and the model can be adapted for everything from auto rickshaws and motorcycle deliveries[75] to laundry.[76]

Conclusion

Chapters 6 and 7 have drawn on diverse examples, from participation and health to education and livelihoods. There are many other use cases I did not address—for example, how place(full) communication enables disaster response or early warning systems.[77] Integrating all these examples, instead of dividing them into domain-specific treatments, I was able to draw focus to two distinct, almost paradoxically juxtaposed general sociotechnical implications of a more mobile Internet: place(less) and place(full) informational exchange. Each is more common in a world with billions of active, data-enabled mobile handsets than in one without.

Nor is the story as universally rosy or linear as this litany of ICT4D examples may suggest. Place(less) and place(full) communication and informational exchange have dark sides, like gossip,[78] terror,[79] coordinated political violence,[80] state surveillance, and tensions between family members in the home.[81] Yet the telephone, too, had "effects in diametrically opposed directions." As Ithiel de sola Pool said about the landline in 1977: "The phone, in short, adds to human freedom, but those who gain freedom can use it however they choose. Rather than containing action in any one direction, the telephone is an agent of effective action in many directions."[82]

This formulation holds for all the communication platforms that came after the phone—the mobile phone, the PC, the Internet—and will hold for the mobile Internet. Yet the centrality of freedom, agency, appropriation, and paradox does not mean that technologies do not shape human behavior. This chapter has outlined how, in these cases, the elements of a more mobile Internet, relative to the fixed Internet and the landline, encourage a range of place(full) interactions with data and other individuals in low-resource settings with a dearth of other telecommunications infrastructures, devices, and users. Like the place(less) interactions described in the previous chapter, these new place(full) interactions present new opportunities for the ICT4D community.

III New Constraints

8   Digital Repertoires and Effective Use

Part III of the book pivots from exploring the potentialities of the shift to a more mobile Internet toward exploring the challenges. As I mentioned when laying out the After Access Lens in chapter 4, the six elements that have fueled the boom in access offer increased opportunities for place(less) and place(full) informational interaction, but also present discernable constraints on engaged, effective Internet use.

I will reintroduce and detail these constraints (and what to do about them) in chapters 9, 10, and 11. However, before I do so, this chapter introduces and merges two perspectives to ground and guide my assertions about what constitutes a "constraint." The first draws on the idea of a *digital repertoire*, from human–computer interaction (HCI) and communication research—allowing theoretical and practical focus on people's choices, skills, improvisations, and adaptations when using a set of technologies, rather than focusing on use of one technology at a time. The second perspective draws on *effective use* from community informatics (CI),[1] providing a way to add valence and directionality to the general idea of the use of digital technologies. For ICT4D, I hope to illustrate how a synthesis of these two perspectives can steer efforts to improve access, use, and socioeconomic outcomes in a mobile-rich, increasingly postaccess environment.

Digital Devices as Tools

The earlier chapters spoke of "access" quite a bit. "Access" to the Internet may be a technically accurate term, however, we would be falling prey to rhetorical slippage if using the term made us think about digital devices as doors or portals that users pass through, unaffected, on their way "to" "the

Internet," or that information passes through untransformed on its way "to" users.

In the era after the invention of the World Wide Web, most people accessing the Internet did so *most of the time* via a personal computer, and spent most of that time in web browsers. (Indeed, people still often conflate "the Internet" and the "World Wide Web" or the "web.") Thanks to the increasingly widespread use of data-enabled handsets, that era has come to a close. Chapters 2 and 3 described how the devices used to access the Internet are more heterogeneous than ever before. Each tool—each *interface*—has its own set of affordances, strengths, and weaknesses. As tools proliferate, it makes sense to compare them and consider what they are good at; that is the goal of part III.

Digital Repertoires

The concept of a digital repertoire is one simple and powerful approach to understanding mobile use in the developing world. By the end of this chapter, the subtleties between mobile-only, mobile-first, and mobile-centric repertoires will be apparent, as will the importance of their distinctions to ICT4D. To be clear, this is not a new perspective. Researchers from HCI and new media research have drawn on multidevice, multimodal theories of digital use for a long time. I am, instead, building on other's work in developing approaches to technology use that account for multidevice, multimodal scenarios, and am suggesting that there has rarely been a better moment to use them.

Mirca Madianou and Daniel Miller offer a particularly useful approach in this vein with their theory of *polymedia*:

Polymedia is an emerging environment of communicative opportunities that functions as an "integrated structure" within which each individual medium is defined in relational terms in the context of all other media. In conditions of polymedia the emphasis shifts from a focus on the qualities of each particular medium as a discrete technology, to an understanding of new media as an environment of affordances. As a consequence the primary concern shifts from an emphasis on the constraints imposed by each medium (often cost related, but also shaped by specific qualities) to an emphasis upon the social and emotional consequences of choosing between these different media.[2]

Madianou and Miller situate their theory as part of a wide array of approaches to "proliferating communication opportunities" by Baym,

Broadbent, Jenkins, Couldry, and many more.[3] However, polymedia stands out as a relatively rare case of a media theory not first developed to describe the practices of prosperous young users in the Global North. Instead, Madianou and Miller draw extensively on ethnographies with Filipino and Caribbean transnational families, many of whom were still facing conditions of relative constraint both in terms of literacy and the cost and availability of communication technologies. They note: "Polymedia is not merely the proliferation of new media and the choices this provides. It is only fully achieved when the decision between media that constitute parts of one environment can no longer be referred back to issues of either access, cost, or media literacy by either of those involved in the act of communication."[4]

Thus, the same formulations that allow Madianou and Miller to capture multidevice, multimedia practices also reveal limits to polymedia's pervasiveness in the present. A different reading of polymedia is that it might be a while before polymedia *is* "fully achieved," because salient differences of "access, cost, and media literacy" will continue to define the choice sets of many resource-constrained users.

Several other perspectives on combinatory and multidevice practices and skills include domestic "mediatopes" (multidevice technical ecosystems in households),[5] "media footprints"/"resource portfolios,"[6] "technology clusters,"[7] "ensembles,"[8] "braided communication,"[9] and "multimodal connectedness"[10]—but I like the relatively straightforward word "repertoire." There are many efforts here, too, including "media repertoire,"[11] "technological repertoire,"[12] "technical repertoire,"[13] "information repertoire,"[14] "genre repertoire,"[15] and "digital network repertoire."[16]

For this text, I use *digital repertoire*, rather than any of the other terms, and particularly rather than "Internet repertoire," because it allows an appropriately broad array of devices and form factors into the equation. For example, a digital repertoire might include devices like USB flash drives, DVD players, digital cameras, or MP3 players that sometimes do not have independent means of accessing the Internet but are still important parts of people's overall practices of interacting with digital content. At the same time, it excludes other nondigital forms of information management and communication, ranging from conventional television, books, and radio to face-to-face conversation. Of course, these all remain important elements in an individual's broader communication or information repertoire, or "communication mode choice,"[17] but are less reliant on interactions with code/services/software created or residing on servers.

The repertoire approach allows the flexible capture and focus on technologies and accompanying skills rather than on the broader contexts around them. In this way, it is much narrower than concepts like *information ecology* ("a system of people, practices, values, and technologies in a particular local environment"),[18] *information system,*[19] *or media/communicative ecologies.*[20] While both repertoire and ecology would allow a look beyond the affordances of a single device, ecological and systemic perspectives necessitate bringing an even wider array of contextual, cultural, and organizational elements into the mix. For example, Taylor and Horst, building on arguments from Bill Maurer,[21] describe their approach to understanding mobile money in Haiti as requiring a wider perspective on "aesthetic ecologies":

> We suggest that people's relationship to mobile money has as much to do with the aesthetics of mobile money as a desired environment and platform that goes far beyond the user interface. The aesthetic ecology of mobile money includes the mobile phone itself, the infrastructure that it depends on, the back-end operations, the agents who sell mobile money services, television and print advertising, and the instructional leaflets that companies give to account holders. . . .
>
> We highlight how instrumental approaches to adoption and appropriation tend to neglect the relationship between aesthetics and practicality that reflects a particularly Haitian cultural logic.[22]

This is an important framing. As the anthropologist Raul Pertierra cautions, "The new media, while invoking its own realities *sui generis*, nevertheless occurs in a broader offline world. The intersection, conflation, or concatenation of these two realities rarely follows predictable paths."[23] Don Slater recommends similarly broad approaches in *New Media, Development and Globalization.*[24] He suggests that communicative ecologies are best approached as comprising communicative assemblages, rather than discrete ICTs, where these assemblages are practices combining technologies with strategies to make the most of them, contingent and specific to the context in question.[25] These "challenge an ICT4D that starts from the taken-for-granted categories of 'media' or 'new media'; or 'ICTs' as given objects (even ones with disputable properties)."[26]

Such critiques of any narrow, instrumental repertoire lens like the one I will employ in this part of the book indicate the unavoidable tradeoff accompanying the choice to emphasize technologies and skills over broader ecologies of technologies, infrastructures, norms, and contexts. Indeed, to understand how Haitians use mobile money, it *is* important to have a conceptual frame that captures the intersections between the technology and the cultural context. Yet once one has populated an ecological frame with data about a given context, it becomes *of* that context. Thus, the After Access Lens in this book evokes only a narrow repertoire concept partly because it is more portable; I can discuss user repertoires that include mobile phones in quite diverse settings, from Cape Town, South Africa, to Chhattisgarh, India, in ways that a broader ecological lens would not allow me to do.

My definition of a digital repertoire, therefore, echoes that of a musician's repertoire—an inventory of his instruments and the skills he has developed to play them. When he moves from city to city, his ecology changes, and his incentives, choices, or decisions to play jazz in one place or rock in the next may change as well, but his repertoire probably changes very little. To convert this to the matter of digital information, I would suggest as a definition: *a person's digital repertoire is comprised of the devices, networks, and services she uses to manipulate digital information, as well as her skills to do so.*

Mobile Phones in Digital Repertoires

Repertoire lenses are particularly useful for assessing mobile telephone use, relative to alternatives. When the mobile phone became popular in the Global North around the turn of the twenty-first century, its early adopters were often individuals or households that already had access to landlines. Social researchers soon formulated explanations of the tradeoffs and choices users made between mobiles and fixed lines, framing them as complements or substitutes depending on the circumstance, the pricing, and the user.[27] In developing countries, meanwhile, research was quick to frame mobile use as a substitute for costly, absent fixed lines.[28] Going beyond economics, some HCI and communication researchers, notably Vincent and Haddon, and Christian Licoppe, used "repertoire" to examine the choices and practices people developed to get the most out of newly available combinations of fixed and mobile telephones.[29]

Today, many users still have access to a combination of fixed and mobile telephones; others have no fixed line, but have more than one mobile line.

For example, Katy Pearce describes how many professionals in Azerbaijan, where she has done fieldwork, carry a smartphone with one SIM for data access, and a second, more basic mobile like a Nokia 1100 as their "talk" phone with a good battery.[30]

The applicability of a digital repertoire approach to scenarios containing PCs and mobiles should be particularly intuitive. Some researchers emphasize a theme that the smartphone threatens to "displace"[31] devices such as PCs, landlines, MP3 players, satellite navigation systems, and digital cameras, depending on the needs of the user, his or her skills and literacies, the context, and the other options available at the time. Others, however, see the shift to mobile Internet (from fixed) as occurring only gradually, instead focusing on complementarities between form factors. In 2010, Nielsen and Fjuk found most users they interviewed in Norway and Hungary saw their mobile devices as augmenting, rather than replacing, their PCs; mobile devices were most appealing when their PCs were out of reach.[32] Lin and his colleagues reported similar findings with a sample of urban East Asian teenagers.[33]

Mobile-Only and Mobile-Centric Digital Repertoires

At this point, all the pieces are in place to introduce one of the central organizing concepts for the remainder of the book. As part I of the book made clear, mobile telephones now far outnumber PCs, tablets, and myriad other connected technologies as the most common means by which people access and use the Internet; this will not change over the next several years. Yet, although most users' digital repertoires include a mobile phone, not all these repertoires are the same. Some users' digital repertoires are "mobile-centric." To them (and quite possibly to you, the reader) the mobile may be the go-to device—universal, personal, task supportive, etc.—yet it is still among two or more choices. In contrast, many other people, particularly in low-resource settings, have digital repertoires that begin and end with their mobile devices; these increasingly common digital repertories are better described as "mobile-only." The key task ahead is to distinguish between "mobile-centric" and "mobile-only" digital repertoires, and, in later chapters, to explore their implications for participation, inclusion, prosperity, and poverty alleviation—outcomes that are each important to ICT4D.

The first nontechnical use of the term "mobile-centric" in the scholarly research literature that I can find is from 1999, when Ciancetta and her colleagues described mobile-centric business models.[34] The first use of the term to describe user behavior was probably by Fox and his colleagues in 2006, referencing emerging youth practices.[35] In 2007, "mobile-centric" may have made its first appearance in the ICT4D literature by researchers at the telecommunications research institute LIRNEasia, who described a mobile-centric "dial-a-Gov" approach to e-government for citizens of developing countries.[36]

Against this background, in 2009–2010 I collaborated on a project with Gary Marsden and Shikoh Gitau at the University of Cape Town. Based on Gitau's fieldwork for her eventual dissertation, we published three papers[37] documenting the emergence of mobile-only modes of Internet use among low-income users in Cape Town. We presented the first of these papers at a workshop entitled "Beyond Voice," organized, not uncoincidentally, by LIRNEasia.[38] A later short paper for the Annual International Conference on Human Factors in Computing Systems in 2010 carried the title of "After Access" shared by this book. As a set, the three papers I wrote with Gitau and Marsden provided many of the initial insights guiding the broader content of this book.

My other inspiration and early in-depth exploration of the role of mobiles in digital repertoires came from a 2009–2011 project with Marion Walton, also at the University of Cape Town. Walton and I contributed a study called "Public Access, Private Mobile"[39] as part of a larger international study of public access to the Internet called the Global Impact Study.[40] It was in this project that I began to work with the digital repertoire concept in more detail. The teenagers we interviewed in Cape Town were *mobile-centric, but not mobile-only.* They had developed complex practices balancing and integrating public access to the Internet via libraries and telecenters with the private (though expensive) access to the Internet they had on their mobile handsets.

As is often the case for social research on new and rapidly diffusing technologies, there were numerous threads coming together in that period, each of which influenced my own work. Walton[41] and Marsden[42] had both conducted other research on mobile Internet use by low-resource populations. Also in South Africa, published work by Chigona and his colleagues,[43] by Ford and Botha,[44] and by Bosch[45] had begun to explore "more than voice" conditions and, as was the case with LIRNEasia, all had published work on mobile Internet use in the Global South before my projects were underway.

Research attention to mobile-centric forms of Internet use has continued to rise. Katy Pearce and her colleagues have worked on differentiating mobile-only and PC-only use, particularly in Armenia.[46] Kumar has explored "non-instrumental" use of "mobile-centric" Facebook users in New Delhi,[47] while Watkins, Kitner, and Mehta use "mobile-only" and "mobile-heavy"[48] to explore emerging forms of Internet use in India. Also in India, my MSR colleague Rangaswamy and her colleagues have documented emerging mobile Internet practices by resource-constrained users in slums of Hyderabad and Mumbai.[49]

Mobile-centric use is not limited to low-resource settings in the developing world. Based on a 2013 telephone survey, Pew Research described how "cell-mostly internet users" in the United States "account for 21 percent of the total cell owner population. Young adults, non-whites, and those with relatively low income and education levels are particularly likely to be cell-mostly internet users."[50] Also in the U.S., Lynn Schofield Clark and her colleagues describe an Internet "Access Rainbow" comprising different levels of use in the home.[51] Sakari Taipale compared the traveling habits (mobilities) of those with fixed-only, mobile-only, and combined fixed/mobile repertoires in Finland in 2011. He found that people in the mobile-only group were of lower socioeconomic status than those in the other groups, and were less likely to drive a car. The study found little support for assertions that mobile-only was a more flexible and free way to conduct a digital life.[52] Other research in the UK compared "mobile" and "non-mobile" Internet users by demographics, finding that those who accessed the Internet from both mobiles and PCs were on average younger, of higher socioeconomic status, and were more frequent Internet users than those with fixed-only. They found very few mobile Internet users who did not have some access to a PC.[53]

Individual users might not describe themselves as mobile-only or mobile-centric, although when prompted, narratives and intuitions about the differences between the modes of use are present. For example, a study with German and American university students revealed a range of lay definitions and conceptions as participants wrestled with the charge of describing the mobile Internet as distinct from the fixed Internet.[54] In the study, Humphreys, Von Pape, and Karnowski found that while overall differences were modest, users described less immersive and more extractive behaviors in the mobile descriptions. Sung and Meyer found differences in the approaches to mobile versus fixed Internet between students in South Korea and the United States. Students in South Korea were more likely to view educational experiences on handsets versus PCs as interchangeable.[55]

Indeed, the terms "mobile-only" and "mobile-centric" do not resonate with everyday usage. Instead, the terms may indicate a lasting bias in the theoretical, technical, and policy communities (particularly in ICT4D) toward "traditional" Internet use—a presumed default experience of Internet use as mediated through a PC browser. Someday the categories may flip and PC use will be the rare case requiring modification, but the challenge is to move beyond these traditional formulations; the research literature on fixed Internet use on a personal computer has a thirty-year head start. Until we can break the habit of modifying "Internet" with "mobile," we should still be careful with what Goggin called the "portmanteau term,"[56] because there are now a variety of Internet experiences, some of which are mediated by the mobile interface.

Enumerating Mobile-Centric and Mobile-Only Use

As noted in chapter 2, without national representative surveys with questions carefully calibrated to delineate different kinds of Internet use, it is difficult to estimate what proportion of users might be mobile-only, mobile-centric, PC-centric, or PC-only in various regions around the world. Even with national surveys, the targets are moving quickly, and large public opinion surveys may not want to spare the real estate for detailed questions about technology ownership and use. For example, for a comprehensive survey by Pew Research, researchers spoke to people in dozens of developing countries about their Internet and mobile use,[57] but only identified *mobile* Internet use on "smartphones," excluding the feature phones I discussed in chapter 2.

There are industry estimates available, but to my knowledge, there is no reliable, universal, replicable, and transparent global measure that tracks the overlaps in the user base between mobile and fixed forms of Internet use, perhaps partially because of the definitional challenges I described earlier. However, some indication of this crossover worldwide can be inferred from reports from Facebook, which in December 2013 reported that 945 million of its 1.2 billion monthly active users accessed their site from a mobile device, while 296 million (almost 25 percent of its total monthly user base) accessed Facebook exclusively from mobile devices like phones and tablets.[58]

There are, of course, limits to the utility of distinguishing between mobile-only and mobile-centric digital repertoires. First, as form factors proliferate, as tablets and phablets and wearables[59] join phones, PCs, and laptops, the lines between PC-based and mobile-based modes of Internet use will become blurrier and sometime not even relevant. At the top end of the market, devices like the proposed Ubuntu Phone[60] and Microsoft's Surface Pro tablet promise "no compromises" approaches merging PC power and mobile form factors. In the meantime, pending broad appeal and affordability, perhaps more people will be able to draw on a smaller set of devices to cover their needs: for example, a tablet and a smartphone replacing a PC, a feature phone, and a landline.[61]

Nevertheless, the distinctions between mobile-centric and mobile-only Internet use will persist long enough to matter for theory about ICT4D and digital inclusion. There are two kinds of mobile-only Internet use: in the first, a user prefers a mobile compared to other available computing options (this is the polymedia explanation); in the second, a mobile is the only practical means of Internet use, due to costs, lack of skills, or other factors (this is the pre-polymedia, constrained digital repertoires explanation). These mobile-only, constrained digital repertoires will likely define Internet use for the majority of the next billion or more resource-constrained users to get online.

Aligning Industry Terms with Research Terms

Beyond the fact that "mobile-only" and "mobile-centric" are helpful shorthand terms to describe user repertoires, and to evaluate potential development impacts, they have the added benefit of aligning closely to the language (and vision) espoused by many in the technology industries. For example, Google and Facebook have each declared themselves to be "mobile first" organizations.[62] However, this detailed discussion of "centric" and "only" reveals a curious vagueness and relativity around the industry zeitgeist of "first." If mobile is first, what is second? Are prognosticators counting devices, proportions of company revenue, or identifying a user experience? Does "first" imply a progression in time in which PC use follows mobile Internet use?[63] (That scenario describes many children in the Global North, who can use their parent's mobile devices long before they set foot in a computer class in elementary school.) Alternatively, as some business writers have already asked,[64] does "first" imply the dominance of the mobile form factor? Or is it the instinct to turn to a mobile device instead of a PC? In the words of one prominent technology blogger, "I now dread using my computer. I want to use a tablet most of the time. And increasingly, I can. I want to use a smartphone all the rest of the time. And I do."[65]

Rather than imposing a hierarchy of devices, strategies, or conditions, an examination of dynamic digital repertoires supports more nuanced and accurate descriptions of use than the term "first" allows. In addition, the idea of a repertoire fits much better with the approach to heterogeneous installed bases described in chapter 4; this reflects the way companies and organizations build software and services that support multiple devices and use cases even if specific users of those systems remain mono-channel.

Skills and Digital Literacy Are Part of Repertoires

Lest we forget, a person's digital repertoire includes not only the devices, networks, and services available in use, but also *the skills and digital literacies* to operate them. Eszter Hargittai's work[66] is illustrative of a large discussion in the research literature about how digital skills and evolving digital literacies[67] matter more than ever in an era when technology use is not only about information seeking, but information evaluating, information creating, and information disseminating, as well.[68] Recent studies with mobile-only or mobile centric users in locations as diverse as Sierra Leone,[69] Sri Lanka,[70] India,[71] South Africa,[72] and the United States[73] point to the ways that gaps in digital literacies, including a lack of imagination and experience about what might be possible with an Internet-enabled mobile device, can prevent people from using all the features on their devices.

Some variant of this gap between possible and actual device use happens to everyone. I could use the devices in my digital repertoire to do many more things than I do, if my digital repertoire also included the knowledge to do so. Yet just as I have had an easy access to devices over the years, I have had years to acquire specific device skills and general digital literacy based on my high socioeconomic status—chances that many others around the world have not had. The distribution (indeed, the reproduction and transmission) of digital skills in a society is reflective of class and power stratifications. Hargittai calls this the "second order divide,"[74] and puts these differences in digital literacy at the center of a rearticulation of the original access-based digital divide concept for a post-access environment. The second order divide will be a feature of the more mobile Internet, as well.

There are other individual factors, general literacy,[75] mental and physical ability,[76] language,[77] emotion and affect,[78] and so on, which of course influence how people interact with technologies, but are not included in the repertoire model. Similarly, the broader cultural, contextual, and social factors I have discussed in this chapter are beyond the scope of the digital repertoire concept. Skills and digital literacies are the only nontechnical components I attach to the concept, for two reasons having to do with their relative malleability, compared to both other individual factors and to broader cultural and structural factors. First, the skills are intertwined with the devices themselves. People's skills can change over time as they work with devices. Second, skills and literacies can be changed through investments in training or through peer learning. More on that in chapter 10.

Effective Use

Focusing on digital repertoires allows us to consider (and evaluate) different combinations of devices and networks, but to which ends? The second half of this chapter takes some steps in this direction, linking ICT use to a variety of possible development outcomes or macrolevel social changes through the idea of effective use.

The "effective use" perspective responds to the reductionism of the digital divide narrative, and stresses how *use* is a more important matter—for study, for practice, for impact—than access alone. For example, in *Technology and Social Inclusion: Rethinking the Digital Divide,* Mark Warschauer highlights the importance of interconnected physical, digital, human, and social resources that both enable and depend on "effective use" of ICTs:[79] "The bottom line is that there is no binary divide and no single overriding factor for determining such a divide. ICT does not exist as an external variable to be injected from the outside to bring about certain results. Rather it is woven in a complex manner into social systems and processes. And, from a policy standpoint, the goal of using ICT with marginalized groups is not to overcome a digital divide but rather to further a process of social inclusion."[80]

Community informatics scholar Michael Gurstein makes similar and contemporaneous points about "effective use":

There is a need now to distinguish between an approach to the "Information Society" and to ICTs which "stresses access/the DD [digital divide]" and one which stresses "effective use." . . . ICTs when used effectively provide significant resources/tools for transforming one's condition—economic, social, political, cultural—whether through obtaining the means for effective use of information and communications capabilities and tools; reaching new markets for small and micro-enterprises; providing the means to bring together dispersed linguistic communities; giving amplification and global voice to unheard minorities (or majorities); for facilitating informed participation in remotely managed political and other decisions; and, for obtaining the interactive services (if remotely) of skilled practitioners.

The key element in all of this is not "access" either to infrastructure or end user terminals (bridging the hardware "divide"). Rather what is significant is having access and then with that access having the knowledge, skills, and supportive organizational and social structures to make effective use of that access and that e-technology to enable social and community objectives.[81]

Many might question the extent to which any external group can define "effective use" on another's behalf. Whether for reasons of pragmatic definitional necessity or due to more fundamental beliefs about the nature of development, autonomy, and agency, Gurstein's own formulation is notably, helpfully flexible, defining effective use as "The capacity and opportunity to successfully integrate ICTs into the accomplishment of self or collaboratively identified goals."[82] Nevertheless, by evoking ideas like microentrepreneurship, civic participation, and information seeking as examples of "effective" use, Gurstein echoes and reaffirms many of the broad perspectives and general frames around what constitutes "development" still common in the ICT4D discourse.

Thus, effective use is directionally progressive and developmental, but broad, contextually dependent, and open to negotiations among stakeholders, observers, policymakers, and especially users. For example, playing games can be effective use, leading to enhanced social capital enhancement[83] or to better digital literacies.[84] So can social-networking site use, which can build social capital,[85] in turn convertible to economic capital.[86] Echoes of these broad formulations appear in research linking ICTs to freedom and agency via the "capabilities approach" pioneered by Amartya Sen[87] and applied to digital technologies by a variety of scholars in ICT4D.[88] For example, Dorothea Kleine builds on the capabilities approach to treat

ICTs as "technologies of choice," which play "a vital crosscutting role in assisting people to lead the kinds of lives *they* value" [italics mine].[89]

Synthesizing Repertoires and Effective Use

By examining the shift to a more mobile Internet from the perspective of digital repertoires, we can understand effective use *relative to what alternatives are available to users*. The perspective tells us that people can use a variety of devices (landline, basic mobile, smartphone, tablet, PC, set-top cable box, etc.) to engage with the Internet, and suggests that no single device allows users to perform all tasks online with equivalent ease. In addition, and particularly relevant for this third part of the book, it allows us to see a difference between the Internet experiences of those who rely on it as their only means of Internet access, and those who can use the mobile Internet as a complement to available PCs or tablets, or both.

Meanwhile, the perspective of "effective use" helps us assess mobile Internet experiences against the prosocial and development goals its users might have. Overall, it is my assertion that on balance, holding all other moderating factors equal, those with mobile-only digital repertoires face greater constraints in effectively accomplishing the full range of possible Internet activities, relative to those with broader digital repertoires. Data-enabled handsets are indeed powerful, accessible, and affordable tools to manipulate digital information, but the digital world does not run exclusively on mobile handsets. To paint a full picture of the shift to a more mobile Internet—and to guide policy and technical investments in the spirit of global socioeconomic development and inclusion—we should consider not only what mobile phones can now do well, but also what they still cannot. In chapters 9, 10, and 11, I will draw on the small research and policy literatures on mobile Internet use in the developing world to highlight three ways in which mobile-centric and particularly mobile-only digital repertoires moderate Internet use, introducing new barriers to effective use of the Internet, even if the exact contours remain vague.

I am not the first to offer reflections of this kind. Marion Walton puts it particularly well, saying, "phones are often designed to slot into a wealthier ecology where physical mobility is easier, phones are individual rather than shared artifacts, computers are ubiquitous, and bandwidth is affordable. These design biases account for some of the creativity that must perforce be displayed by African users (and indeed by low-income users do around the world) as they reinterpret and appropriate handsets and networks."[90] Later, drawing on Paul Leonardi, she adds: "intangible infrastructure such as mobile handset features, SNS architecture, network speed and tariffs all constitute specific 'digital materialities' which play a role in shaping participation."[91]

In *A Human Rights Approach to the Mobile Internet*, Lisa Horner lists affordability, usability, content creation, generativity, gender norms, and specific cultural contexts among the major "challenges for advancing human rights and empowerment through the mobile internet."[92]

Phillip Napoli and Jonathan Obar review scores of articles to compile a list of concerns that include technological capabilities; memory, speed, and storage; content availability; network/platform architecture; usage patterns; information seeking; content creation; and skill sets across platforms. They state: "While mobile Internet access may address the basic issue of getting individuals who previously did not have any form of Internet access online, the differences between mobile and PC-based forms of Internet access can reinforce and perhaps even exacerbate, inequities in digital skill sets, online participation, and content creation. Consequently, mobile only Internet users become, in many ways, second-class citizens online."[93]

Writing about local content, Mark Surman, Corina Gardner, and David Ascher note "the mobile internet we have built so far looks nothing like the wide open, come-as-you-are, read-write world of HTML and traditional online publishing platforms. In many ways, the mobile internet is 'read only' not just because authoring content is difficult on small screens but because mobile content—media, apps, and services—are distributed through much more restrictive channels than the early web, or even Web 2.0."[94]

Writing about the United States, Astra Taylor in turn draws on Susan Crawford:

"While we still talk about 'the' internet, we increasingly have two separate access marketplaces; high speed wired and second class wireless," law professor Susan Crawford explained in an op-ed. "High-speed access is a superhighway for those who can afford it, while racial minorities and poorer and rural Americans must make do with a bike path." Mobile connections cost more, are subject to data caps, and are less open, adaptable, and generative. Handheld devices simply can't compare to personal computers if you want to do a long-distance learning program, fill out a resume, start a business, program an app, write a long essay, or edit a feature length film.[95]

Even the ITU offers an opinion on mobile-only Internet use:

It is also important to note that while mobile-broadband technology helps to increase coverage and offer mobility, the mobile networks and services currently in place usually only allow limited data access, at lower speeds, which often makes mobile-broadband subscriptions unsuitable for intensive users, such as businesses and institutions. High-speed, reliable broadband access is particularly important for the delivery of vital public services, such as those related to education, health and government. The potential and benefit of mobile-broadband services is therefore constrained when mobile broadband is used to replace, rather than complement, fixed (wired)-broadband access.[96]

By drawing across these and other[97] critiques, I can continue with the book's task of presenting mobile-only digital repertoires as a set of overlapping and sometimes contradictory affordances and experiences, rather than as a singular artifact. To reconcile and expand these observations, I introduce three broad areas as the final part of the After Access Lens: *metered mindsets, challenging production scenarios,* and *circumscribed structural roles.* In aggregate, they hold keys to understanding why, although the user of the $40 feature phone and the $1,000 PC can both "access" the Internet, ICT4D theory and practice can no longer assume they are using the *same* Internet, to the same effects.[98]

Part II framed place(less) and place(full) informational activities as potential new avenues for empowerment, allowing mobile Internet users to do new things. The tone of part III is admittedly less upbeat, framing some of the same core elements of mobile Internet technologies as potential constraints, allowing mobile Internet users to do fewer things, less effectively. As you move through the remaining chapters in part III, keep in mind how the digital repertoire perspective is required before this shift to consider constraints makes any sense. Those with broad repertoires, with regular and affordable access to PCs and to large amounts of reasonably priced, reasonably fast data, are unlikely to be particularly constrained by their mobile devices. If a task is too difficult or annoying or not possible on a mobile device, on a pay-as-you-go metered connection, or on a 2G signal, a user with a broad repertoire can draw on other elements in the digital networks. A dozen other factors from literacy to language to social structure to habit might still discourage the effective completion of that goal, but it will not be the technology's fault. Instead, individuals most likely to be additionally constrained by mobile interfaces are those with no practical options to access and use the Internet other than the mobile antipode. Members of the research, policy, and practice communities worried about ICT4D and broad-based inclusion in the information economy should be most worried about individuals with mobile-only digital repertoires because they are the users with the most compromised digital experiences.

In chapter 2, I detailed ongoing efforts to improve access. The chapters that follow use the After Access Lens to surface a separate set of possible technical interventions and policy priorities. The recommendations from here forward only make sense if one treats effective use, rather than mere access, as a central outcome of interest and as a goal worth investing in. It only makes sense to work on the mobile-specific barriers to mediated social and economic inclusion, participation, and productivity outlined by Gurstein if (a) one feels they can be somewhat mitigated by technical and structural changes and (b) that this mitigation is a worthwhile goal meriting attention by some combination of market, public, and third-sector actions. I will reflect further on the need for further policy evidence and theory refinement on these matters in concluding chapter 12. In the meantime, the rest of part III identifies actions that practitioners, policymakers, and particularly technologists could pursue in order to accelerate and expand effective use of the Internet by those with mobile-only digital repertoires, with an eye toward the kinds of outcomes of interest to the ICT4D community.

9   A Metered Mindset

In 2010, Marion Walton and I contributed to the Global Impact Study, coordinated by the University of Washington with support from the Bill and Melinda Gates Foundation and Canada's International Development Research Centre (IDRC). Our study, entitled *Public Internet, Private Mobile*, asked a deceptively simple question of teenagers we found in libraries around low-income neighborhoods in Cape Town: "Your phone has Internet—why are you at a library PC?"[1]

We drew data from four interconnected activities: semistructured interviews with library and telecenter operators around Cape Town; open-ended interviews with users; observation and task analysis; and a brainstorming session with stakeholders in the public access community in Cape Town. From the title and theme of the work, readers can anticipate that the digital repertoire concept figured prominently in our minds. We looked specifically at the interplay between modes of Internet use, and explored the practices teenagers had developed to help them negotiate the respective strengths and weaknesses of their technology options. I will come back to the issue of "homework" and "finding work" in chapter 10. For now, I want to talk about costs.

The young people we spoke to had data-enabled mobile phones (mostly feature phones), but were in the public library in part to take advantage of the half hour of "free" data they were allotted each day. Quite simply, the subsidized (free) access available for fifteen or thirty minutes a day via the libraries made walking to one and waiting in line for a terminal worthwhile. One library manager explained: "People come to the library [be]cause the Internet is free, even phones are too expensive. We've even had clients saying that they do have Internet at home, but they prefer to work at the library [be]cause they don't have to pay for the data."[2]

Metered Mobile Billing

Chapter 2 described the pay-as-you-go/prepay model for billing cellular airtime, which has now become the mechanism for how the majority of mobile users in the Global South pay for data. Pay-as-you-go pricing, often sold via scratch cards by corner shops and street vendors or in electronic kiosks, helped create a fluid airtime economy. Requiring no credit check or monthly billing, the model created opportunities to enroll hundreds of millions of new users who otherwise would be too costly for mobile network operators to serve,[3] and fit well with the way many households liked to manage their expenditures, incrementally, bit by bit.[4]

MNOs tend to offer postpaid cellular subscribers data *plans*. These plans generally allow a set amount of data consumption per month, after which consumption is capped (shut off) or throttled (slowed to a crawl). As prepay cellular customers began to demand data, MNOs responded by adapting pay-as-you-go billing for data, allowing prepaid users to convert airtime to bits of data whenever they clicked on a link or opened an app. Other prepaid users get better rates by purchasing blocks of data (from ten megabytes up to one gigabyte or more) to use within a set amount of time.

Thus, whether directly deducted from airtime balances, purchased in a block for consumption, or consumed in a monthly allocation with a "cap,"[5] mobile data offered by MNOs is *metered*. These metered approaches, also called "usage-based pricing," are so central to the economics of delivering data over cellular/wireless networks that I listed it as one of the six core mobile elements that helped drive the boom in mobile device adoption and use.

Yet usage-based pricing has a downside; it creates a "metered mindset" among people, who remain aware of the incremental costs of using their devices. The deterrents to effective use are not simply the cost of data, but also how data is priced, how it is purchased, how it is perceived, and how it is experienced. I want to spend a full chapter on this issue not because I wish to rail against charging for data, or to dispute the need to use pricing as a way to manage capacity on wireless networks, but rather because I think the metered mindset has not been sufficiently addressed in the research and practice literature on mobile Internet use. That the metaphors "surfing the Internet" and "browsing the World Wide Web" are integral to our shared understanding of an Internet experience is a testament to this gap. As hundreds of millions, perhaps billons, of new users join the Internet via mobile devices, in mobile-only or mobile-centric digital repertoires, they are unlikely to surf and browse; rather, they will dip and sip.

Starting with mobile airtime/voice calls, we can look to what Taylor calls "prepay literacies"[6] for illustrations of how the metered mindset influences mobile use. Some users will maintain two or more prepay accounts, swapping SIM cards or jumping back and forth between networks. (That is why there are dual-SIM phones for sale through the developing world.)[7] Other people use intentional missed calls to send messages to each other to avoid incurring the cost of a phone call. Missed calls are a language emerging out of constraint; a single missed call can mean "call me back" or "I'm thinking of you" or "bring home some bread" depending on the context of the moment. Two or more missed calls placed in succession might mean something else, if the parties have agreed in advance.[8] There are successful businesses in India, like ZipDial, fielding literally hundreds of millions of missed calls, on behalf of businesses and organizations that are happy to call prospective customers right back.[9] The CGNet Swara IVR I discussed in chapter 3 uses a missed call to initiate a session, so that its users will not have to pay. In both the ZipDial and CGNet Swara examples, the target users are familiar with how missed calls function, and know how to use them to save money; in turn, the systems are specifically designed to be responsive to their users' metered mindsets.

What is important for this chapter are the ways in which practices that started with voice (airtime) are moving over to govern the consumption of data (megabytes), as well. In the early days of mobile Internet access and use, unlimited mobile data was available to monthly subscribers on some networks in the Global North. Someone could pay for data services and get email, download ring tones, and browse the early mobile web without straining the networks of operators. (Even today, in developed markets, some smartphone users have unlimited data plans.)[10] Of course, many others could not afford an unlimited plan, and early research on youth adoption of mobile Internet in the UK suggested that *cost* was a major barrier to ongoing use.[11]

Legacy plans notwithstanding, the iPhone and its smartphone contemporaries changed the prospects for unlimited mobile data.[12] MNOs found their systems under siege in tech-savvy cities. Smartphones became data-hungry, gobbling up network capacity faster than network operators could serve it up. In response, MNOs in the Global North began to offer postpaid data plans with a few gigabytes of data included per month, after which a user's data access was "capped"—blocked or slowed to a crawl. A similar scenario unfolded with domestic/residential cable service in the United States, where companies like Comcast began to withdraw previously unlimited flat-rate plans in favor of subscriptions with data caps (measured, mercifully, in the hundreds of gigabytes) in order to rein in the most voracious of video downloaders and file sharers.[13]

Experiencing Usage-Based Pricing

Yet aside from those early adopters, grandfathered into unlimited mobile data plans, living in Japan or South Korea, or part of corporate mobile plans, *most* global users will experience usage-based pricing for mobile data, either as a pay-as-you-go per-byte charge, as a small (bundle) of megabytes, or as a monthly plan with a capped amount of data allowed. Thus, one cannot understand mobile Internet use without understanding usage-based pricing.

For Many, Mobile Data Is Expensive . . .

One way that users understand usage-based pricing is as an overall expenditure, relative to their household incomes. The Broadband Commission for Digital Development, a joint initiative of the ITU and UNESCO, has been working since 2010 to promote accessible, affordable broadband access worldwide. They suggest setting a target for entry-level fixed broadband of less than 5 percent of average monthly income, a figure most countries in the developed world have attained.[14] Fixed broadband is less affordable (and available) in many developing economies, but the affordability challenge for mobile broadband is also difficult. The ITU estimates that in 2013, 500MB of prepaid data used on a mobile handset cost 1.3 percent of monthly gross national income (GNI) per capita in developed economies. In developing economies, the same 500MB would cost 15.7 percent of GNI per capita. Africa is particularly constrained (both by frequently higher tariffs and frequently lower average incomes); that same data bundle would represent 38.8 percent of GNI per capita.[15] The outlays on telecommunication services are smaller in absolute terms in the Global South than in the Global North (the average monthly revenue to MNOs per user in

Africa in 2012 was less than US$8, versus nearly $50 in North America for the same period[16]), but may represent a larger (and more painful) share of disposable income. Although mobile and fixed bandwidth costs are declining worldwide,[17] they will remain palpably high from the perspective of those below median income levels.[18]

. . . and Hard to Understand

Data pricing is more complex than building a bundle of minutes.[19] One study aggregated prepay mobile data tariffs across twenty-seven African nations in 2012. The study found a range of daily (even hourly) plans: 5–20MB plans, weekly plans, and monthly plans, with costs for data varying widely from US$.03 for 1MB of data in Ghana to US$4.69 for 50MB in Libya.[20] This jumble of plans and offers (some much better than others, on a per-bit basis), make it difficult for users to track their time and consumption.[21] The ITU notes that in Cameroon, MNOs offer Internet plans that last fifteen minutes![22] However, these fragmented plans can also be seen as evidence of competition, and of providing consumers with choice. In some markets with strong competition and low-priced spectrum from national regulators, price wars may drive down data prices for many who are attentive to (and can understand) the deals.[23]

However, there is an insidious dual influence at work. In addition to the actual constraints of per-byte pricing, there are experiential constraints associated with the lack of transparency about units and experiences. With voice calls and SMS messages, on the one hand, users can pay by the minute or the message, and have a sense of how their behavior maps to expenses. With data, on the other hand, prepay users purchase some aggregation of mysterious bytes. Some of these bytes are assembled into larger, more recognizable increments like songs or photos; other bytes are consumed in the background, evaporating thanks to chatty background processes.

As part of a study on bandwidth constraint, my colleagues and I asked members of thirty-two South African families with metered Internet connections at home if they could tell us how many megabytes a 3.5-minute music video—Shakira's *Waka-Waka*, in honor of the 2010 football (soccer) World Cup—would consume on YouTube. Alternatively, we asked how many times they could watch *Waka-Waka* on their home PC with a monthly allotment of 1GB. The answers varied tremendously, from 2 to 4,000MB (median = 15MB) and from 2 to 500 times for 1GB (median = 20 viewings). The correct answer, assuming a standard-quality YouTube video at 300 Kbps consumes 2.25MB per minute, is that a 3.5-minute video "cost" 7.8MB and therefore one could watch that video 127 times before reaching the 1GB cap.[24]

So People Consume Less Data than They Would Like To . . .

Under metered conditions, the experience for many people who are connecting to the Internet over the mobile/cellular network is not one of abundance, but of looming, ambiguous, persistent constraint. How much would your browsing change if you knew that every new click—every new page request or download—cost you money? If every byte drains a user's airtime, not to mention his battery,[25] or if every few pages he browses eats up his "top up" data bundle, there are a set of rational reactions. Ericsson's study suggests that mobile users in the UK and United States with unlimited rate plans consume twice as much video per month as users with limited plans; in Japan, the study reports, those with unlimited plans consumed ten times the video as those with limited plans.[26]

Reactions under metered pricing or caps may not be economically efficient. For example, in a paper entitled "Cognitive Bias in Network Services," Stanojevic, Erramilli, and Papagiannaki conducted a study using the traffic logs of hundreds of thousands of users of a European carrier. They found that people had difficulty selecting optimal bundles of voice, SMS, and data tariffs. Many ended up overpaying by an average of 25 percent, in order to insure themselves against exceeding their usage caps. More important for this case, a secondary finding identified those customers who moved from pay-as-you-go data tariffs to bundles with a set number of minutes of talk time, plus unlimited SMS and data. Voice and SMS traffic under those bundled purchases doubled; traffic on the 3G network increased *forty-seven fold*.[27]

Costs matter so much that people are willing to make tradeoffs on the quality of their data use. My colleagues and I did a study[28] to explore how users might trade off costs and quality in a video that we had degraded to the point of visible losses to fidelity. In experimental settings, we manipulated the video quality and communicated to users how much they might save buying the inferior, copied video instead of a higher-quality original video. Results in the field in low-income communities in India are still preliminary, but suggest that, among our (nonrandom) sample, one-third of respondents would select noticeably degraded videos if they could get more of them for the same lower price.

In the capped DSL study discussed earlier, we found that "when the meter is running," families have to make decisions about how to allocate bytes. They must balance Facebook versus work, determine how much data to give to kids, and how much time or money to spend on the Internet to avoid running out of credit before the end of the month.[29] Self-censoring leaves bits on the table, while gorging at the end of the month on "leftovers" means not doing things earlier in the month that you might want to do. Even if the cap is managed correctly, prohibitions on certain *types* of use or family limits on Internet time, rather than on specific calculations about the value of an Internet session, mean that there is stress, uncertainty, and inefficiency in the mix.

. . . and Seek Alternatives

In response to these pricing constraints—real and perceived—many people try to replace bytes consumed on the mobile networks with bytes moved by other means. In places where Wi-Fi is available, people "offload" a lot to public or private Wi-Fi networks. For example, apps like Dropbox or Google Drive have default settings to only transfer files when on a Wi-Fi connection. Three studies place the estimate of total mobile web connections coming over Wi-Fi, rather than cellular networks, at between 30 and 80 percent.[30] (This includes Wi-Fi connections at home, through a cable modem or DSL connection.) However, Wi-Fi (especially free/open Wi-Fi) is scarce in much of the rest of the world, and many mobile-only Internet users will not have the option of a dependable Wi-Fi connection, especially an unmetered one.

A second approach involves offline sharing, trading content from phone to phone not by sending files over the network but rather by proximate sharing through exchanging memory cards, wireless Bluetooth transfer, and visiting "download shops." In cities and towns throughout the developing world, these shops are businesses that specialize in selling pirated content for local transfer to customers' phones.[31] Sometimes content is purchased in bulk, by the gigabyte instead of by the title.

A third approach would be to use a different combination of tools and networks in the digital repertoire—if not a personal PC then perhaps a shared one at a cybercafé or telecenter, like the teenagers in the study I described at the beginning of this chapter.

Those with a mix of devices and the wherewithal to afford a mix of connectivity options may choose multiple configurations at home, subscribing to a household fixed connection and paying for cellular service; others will elect to go cellular only. The GSMA estimates that the break-even point for those who want a single source for Internet data maybe about 5GB of consumption a month; below 5GB, the "usage based" pricing of many mobile plans makes sense. After 5GB, and particularly at 12–18GB (a figure very easy to reach once video is in the mix), fixed connections are more cost-effective.[32] In other words, mobile-only users consuming more than 5GB a month are getting a bad deal, relative to what they could get if they could also have a fixed Internet connection.

Design for the Metered Mindset

Given the behaviors we can observe under conditions of usage-based pricing, there are a variety of things practitioners, policymakers, and technologists in particular can do in addition to continuing to push downward on the average price of data, and making apps, content, and devices more data efficient (as I had described in chapter 2). In the rest of this chapter, I outline a few ways in which those seeking to reach or engage with mobile-only Internet users can empathize with the metered mindset, designing for it where possible.

Increase Visibility and Control of Data Use

One way to improve user experiences under usage-based pricing scenarios is to ensure that users have a clear idea of how many bits they are using, and why.[33] Building on the idea of prepay literacies,[34] and falling under the umbrella of a repertoire approach, users need to have an understanding of the costs associated with their actual behaviors. Thus, the emergence of tools to more effectively and precisely monitor bandwidth consumption on mobile devices is encouraging. By 2012, third-party bandwidth-monitoring applications had appeared in the application marketplaces for Android, BlackBerry, and iOS, while newer versions of operating systems offer bandwidth monitoring integrated directly into them.[35] These examples illustrate that the market is becoming attuned to the needs of bandwidth-constrained mobile users. Innovations will surely continue.

In a pilot study in Ghana, researchers at Google adapted a smartphone browser in a few ways: indicating per-page download costs by tracking credit balances, alerting users at risk of overspending, and allowing easier data top-ups direct from the phone. An experiment found that, compared to users in a control condition, those with the experimental software installed went online more frequently and spent less money while there—an increase in overall efficiency of use.[36] Similarly, Chava and her colleagues[37] propose a "cost aware mobile-web browsing mechanism" that would allow users to configure their mobile browsers to degrade/compress downloaded content dynamically—relative to the amount of data they have left in their accounts.

Pilot studies of this kind indicate how further gains in specificity, transparency, and clarity can be gained. My research with capped DSL users in South Africa,[38] and Sambasivan and colleagues' work with Google's Smartbrowse pilot in Ghana,[39] suggests that there might be additional benefits in representing data consumption *in currency units, rather than simply in bytes.* In the case of Google's Smartbrowse, showing an alert on the estimated cost of a page about to be loaded resulted in hundreds of declined pages (meaning, pages the users in the trial elected not to load). The more expensive the page, the more likely a user was to decline to load it.[40]

This finding from an experimental setting—showing users how much pages cost to load—will be challenging to deploy in the real world because providers can make good returns by offering a wide variety of rate bundles to compete on convenience rather than on price.[41] Over time, if a user tops up with more than one bundle, the average cost of all the bytes stored in his account can rise or fall. Operators and software providers (whether operating system designers or third-party app makers) may have to work together to capture and process the data necessary for a user's phone to communicate data consumption consistently and reliably in currency, rather than mysterious bytes.

Support Intermittent Connections

Other frugal engineering efforts are working on improving the reliability and utility of services under conditions in which the connection comes and goes. In places where Wi-Fi is common, mobile users can schedule down-loading so that bandwidth-intensive tasks wait for less expensive (or faster) connections. However, in resource-constrained settings where the cell net-work may be the only Internet connection, this is less practical. Instead of toggling between more and less advantageous connections, mobile-only users may be toggling between a slow, expensive connection and no con-nection at all.[42] The intermittency of mobile data connections can be due to changes in signal quality and congestion, movement in and out of zones with coverage, battery life/access to electricity (a huge barrier to effective use),[43] or, of course, airtime balances.[44]

Designing for intermittent connections may mean *reimagining data-hungry elements of cloud services*. Walton explains: "even as smartphone and feature phone handsets increasingly emphasize cloud-based sharing of media, high prepaid data costs mean that these 'affordances' can be unaf-fordable."[45] For users with intermittent connections, downloading may still trump streaming, and local replication, including peer-to-peer-sharing, whether legal or not,[46] may trump online backups and synchronization to the cloud. For example, a 2014 version of the Facebook app on iOS and Android had extensive offline capabilities where one could like and com-ment upon posts, photos, and pages while offline and once online, all of the likes and comments uploaded to Facebook.[47]

There may also continue to be a role for community-access hubs hous-ing "heavy" audiovisual content that can be exchanged with users' hand-sets.[48] As I mentioned earlier, download or "cyber" shops remain popular in some markets, in which audio and video content, often pirated, is loaded on to users' phones via SD cards, bypassing download charges (as well as royalties and content charges).[49] There may be ways to bring that practice out of the shadows to be a force for legal, widespread, affordable propagation of content: for example, to distribute educational content across devices without using costly cellular connections.[50] To return to the study with the Cape Town teenagers I began with, the availability of unme-tered free Internet access in public venues transformed the repertoires of the teenagers from mobile-only to mobile-centric, and at least partially assuaged the feelings of cost constraint limiting their effective use of the Internet. Community access hubs could play a similar role on a larger scale, even in an era when most adults have data-enabled mobile phones of their own.

Subsidize Data Charges

What if telecenters and libraries were not only places to get free data? Perhaps the most direct way to address the metered mindset is to stop charging users for some or all of the bits they consume, shifting the cost of Internet use from users to MNOs, service providers, or other sponsors.

Some forms of this approach involve promoting different prices for different times of day, reflecting different demand on networks. Since fewer customers are using their networks, for example late at night, MNOs are offering bundles that make those bits cheaper and incentivize use at those times.[51] Well-informed customers can either stay up late or schedule data-intensive tasks for these cheaper times of day. These differential pricing schedules could be made dynamic through software innovations, adjusting much more quickly to changes in network load, rather than only according to predetermined schedules.[52] Pricing can also take the form of subsidies and MNOs can elect to give away bundles of data free of charge. For example, Nokia offered free data to its Lumia Smartphone buyers in India in 2013; it had to work directly with MNOs to make that offer.[53]

In general, there is certainly some value in trying to find ways to make pricing more flexible, not only between users and their MNOs, but between users and other users. For example, airtime transfer systems active on many mobile networks allow users to gift airtime credits from one account to another, and have been used successfully as elements of compensation and promotional schemes.[54] There appears to be room to create systems for gifting or transferring data credits, as well.[55] For example, in some small enterprises in India, owners are expected to pick up the cost of their employees' business-related mobile use, even if everybody is on individual prepaid plans.[56] How might that kind of informality be recognized by those deploying mobile data services? As more prepay users are expected to interface with their organizations, employers, schools, and governments by the mobile Internet, do we know how to compensate people for the bytes they use on someone's behest?

Yet as Cave and Mfuh[57] point out, there is no equivalent to "calling party pays" for data on cellular networks. It costs one user to post a photo to her blog from her phone, and costs each of her friends to look at that photo on the blog. To echo chapter 5, one person can pay for the call to establish a switched connection, however both (or all) will pay for the interactions with the server.

Cue the hunt for a toll-free approach to data tariffs—an "800 number" for data, in the words of a U.S.-based AT&T executive.[58] It is not at all surprising that by 2013, a patchwork of websites (like Wikipedia and Facebook) had struck a variety of agreements with specific operators in specific countries to "zero rate" [59] (cover the cost of) visiting specific sites.

Four prominent institutions—Google, Facebook, Twitter, and Wikipedia—have pursued the zero-rating strategy most publically. Working with mobile network operators in dozens of developing countries, Facebook introduced 0.facebook.com ("Facebook Zero") in 2010.[60] At the time it was a simple site without all the games or photo uploading of full-blown Facebook clients; even photos were placed "a click away," and viewing them would incur data tariffs. However, as the name suggests, it is free, and can be accessed from any handset with a mobile browser, albeit in the specific markets and on the specific MNO networks with which Facebook had negotiated deals. It is thus appealing to resource-constrained users who are becoming a major source for Facebook's growth.[61] Similarly, the "Facebook on every phone" promotion ran in cooperation with operators in India, Russia, Pakistan, Bangladesh, Turkey, the Philippines, and Brazil, as well as the UK and Germany in 2012, offering ninety days of free data for users accessing Facebook through its J2ME (feature phone) app, which Facebook claimed worked on over 2,500 different handset models.[62] Facebook also subsidized data rates for its messenger service, working with eighteen carriers in fourteen countries.[63] Google offers similar zero-rated promotions in selected markets via its "Google Free Zone."[64] Twitter offers "Twitter access" to MNOs in several developing countries.[65]

Two examples are particularly germane to the ICT4D conversation, since they link explicitly to prosocial or development aims. In 2014, Wikipedia rolled out "Wikipedia Zero" to MNOs serving 330 million users in thirty-six countries.[66] Also in 2014, Internet.org, an industry partnership led by Facebook, interested in expanding Internet access in the developing world, launched a zero-rated Internet.org app in Kenya, Zambia, and Tanzania; in 2015, it added Colombia[67] and India,[68] which I will discuss in chapter 11. In each country, a mix of other content and services considered relevant to development (crop prices, maternal care information, etc.) was bundled alongside Facebook in the zero-rated Internet.org app.

The encouraging news across all these initiatives suggests that the user-pays-for-every-byte model is not the only model available. In certain situations, a user has access to more content and functionality at lower cost than she would have had prior to the rollout of zero-rating agreements. By 2014, zero rating had spread to music streaming, and was making headway both in the United States and in other geographies. In the United States, MNO AT&T had "sponsored data" arrangements with ten companies. One of those companies in turn "launched its own sponsored content store and a development toolkit that lets Android and iOS developers integrate sponsored data into apps" while another "plans to give users of its mobile office software sponsored access to file access and transfers."[69] These agreements, in the short term, help shift the burden off individual users and encourage interaction with these sites, but the agreements are far from universal, and point to strains in the pricing system as a whole. In this chapter, I stressed the positive, narrow case for subsidizing tariffs as a response to the metered mindset. In chapter 11, referencing Internet.org in India in particular, I will return to the broader questions concerning power and control, and the implications for individual users and for the Internet as a whole if zero-rating becomes more widespread.

Conclusion

This chapter explored the metered mindset, and described how "pay as you go" and "usage-based pricing" made the jump from mobile voice to mobile data. It surfaces a tension: the *good*—rapid deployment of a mobile network around the world, accessible to people who lack the account or credit ratings to support subscription plans—may be at odds with a vision of the *perfect*—of everybody on the planet having the same levels of read-write access to the networks themselves.

From video on demand to critical health or government services, nearly every experience delivered via the mobile Internet still costs most users "by the bit." Yet resource-constrained users, mobile-centric and mobile-only users coming online do not browse and surf, they dip and sip, carefully conserving airtime and the balance on their data bundles. Thus the metered mindset itself is a drag on effective, engaged use of the Internet.

This drag on effective use should be important to those who would declare the "digital divide" closed thanks to the spread of a more mobile Internet, and to those who want to use new mobile Internet approaches specifically for prosocial development interventions. For example, it is conceivable that schools in resource-constrained settings could make up for a lack of physical textbooks by recommending that students do research on their mobile devices,[70] but is it good policy or pedagogy to require families to pay by the megabyte to do their homework? Similarly, if the highest-quality, most informative, and transformative video outlining new and life-saving prenatal-care behaviors counts against a daily or monthly data cap, how can an NGO ensure it will be viewed by the medical practitioners who need it?

The metered mindset will persist, because despite good efforts to compress, cache, and conserve, an increasing array of Internet experiences are bandwidth hogs. Consider the enthusiasm for streaming-video services, remote backups, and cloud-based editing software (among others). A lot of the innovation in the Internet space originates from people and organizations in places with copious, unmetered bandwidth, for users they presume have copious, unmetered bandwidth. Will the coming wave of mobile-only Internet users in the Global South be able to use these services effectively? Even if their handsets are capable of accessing broadband content, their wallets will not support the activity.

This discussion of the metered mindset adds another layer to the gradations discussed in chapter 3. We need to remain cognizant of which "Internet" we are discussing in theory and policy circles. Are we discussing copious bandwidth, heavy content, and big screens, or are we talking about a lowest-common-denominator Internet, accessible only intermittently, for a couple of dollars a month?

The goal of this chapter is not to paint a gloomy scenario or decry that bytes cost money. The goal, instead, is to introduce some additional caution into the utopian rhetoric of a closed digital divide. This matter of a pay-as-you-go Internet, harkening back to the earliest days of dialup, demands sensitivity and further attention from the research, policy, and technology communities alike.

10 Restricted Production Scenarios

Have you ever waited until you could get back in front of a PC to write a long email, instead of using your smartphone? If so, that pause is evidence of another kind of constraint confronted by some with mobile-only (and to a lesser extent, mobile-centric) digital repertoires. Some of the same elements and affordances that make a data-enabled mobile phone so compelling as a complement to a PC—as a portable, personal, always-on device for coordination and consumption—can at times make it a poor substitute for a PC. This chapter will explore how some persistent limits on certain kinds of "production scenarios" on mobiles should be of concern to those interested in ICT4D and digital inclusion.

In chapter 9, I presented the example of teenagers in Cape Town, visiting computer centers in libraries in the afternoon to use the Internet, instead of using the data capabilities on their phones.[1] Although "free" was a big draw, the increased productive affordances provided by the PCs were equally important for doing their homework. These productive affordances are the topic of this chapter.

The teenagers' homework often involved reports, requiring an array of informational activities—searching for information, assessing source credibility, capturing and sorting material, converting to essays—that remains difficult on their feature phones. Not impossible, but *relatively* costly and *relatively* frustrating, compared to PCs. The bigger screens, multitasking, keyboards, and mouse pointing devices available at the libraries all made the process of doing their homework much easier and more efficient. In addition, printers made the process of saving information possible. Since they were limited to thirty-minute blocks of time, students used what one of them called "smash and grab" techniques to get in, search a bunch of sites, and print out all the information they would need to later write their homework assignments. Note they did not save sites or materials up to personal storage accounts in the cloud, since retrieving them later would cost money. Nor did they transfer files locally to their phones, because the libraries did not allow phone-to-PC connections, for fear of viruses.

Framing Productivity and Production

Recall the list of Apple's top ten downloaded free iPhone Apps from chapter 4—it contained four games, plus YouTube, Google Maps, Snapchat, Instagram, Skype, and Facebook.[2] Not a single app from the "business" or "productivity" categories on the iOS app store cracked the top ten for popularity. Similarly, the Flurry study I referenced in chapter 4[3] indicates that phone calls, games, messaging, social networking, and a bit of photo sharing probably account for the bulk of most users' time on data-enabled devices. Making sense of these figures requires caution. In focusing on production scenarios, I am inviting frames that privilege some forms of informational activities over others, and I may seem to be judging the mix of behaviors and activities people pursue via their mobile devices. Here are four considerations to keep in mind as I discuss production scenarios.

First: not everyone needs or wants to be an information producer. Not everyone's livelihood or passion depends on manipulating information stored on servers via mobile devices. A "development narrative," to use Slater's term, that situates every person as a knowledge producer in a global information economy is a dangerously oversimplified narrative.[4] Although information creation and manipulation is more important to a wider range of livelihood activities than ever before, only an oversimplified narrative would expect, predict, or demand that *everyone* download an office suite, video editor, or website creator to their handsets.

Second: a mix of activities can be information production. What is a productivity tool, anyway? Many small businesses and individuals can engage in information production and manipulation via multipurpose social media instead of dedicated productivity tools. Susan Wyche and her colleagues, for example, depict how individuals in Kenya use Facebook for "the hustle" of running and promoting their microenterprises.[5] These microentrepreneurs have not registered their own dedicated professional pages, or used any special software. They are just communicating, coordinating, *and* producing information (albeit on a small scale) using general-purpose tools, the same tools that happen to be at the core of their everyday digital lives.

More broadly, the Internet, and particularly now a more mobile Internet, has welcomed well over a billion people onto social media platforms where many write posts and status updates, many others take and share images, and some share videos. In so doing, all are engaging in some manner of artistic, creative, generative, expressive digital practice. Mary Meeker suggests that by 2014, users collectively were uploading 1.8 billion photos *per day* onto Facebook, Snapchat, Instagram, and at WhatsApp alone.[6] These practices—what Yochai Benkler calls "peer production" and "social production"[7]—are the zeitgeist of the age. Thus, the Internet has certainly enabled broad-based participation in the creation of cultural and economic digital artifacts. From Facebook to freelancing, the themes of digital participation and production—via terms like "architectures of participation,"[8] "produsage,"[9] and "prosumption"[10]—have come to characterize the emergence of new structures of work and contribution. In some very important ways, mobiles have invigorated rather than stymied digital "production."

Third: everyone uses their phone for a mix of instrumental and noninstrumental purposes. In 2004, I led a small study interviewing hundreds of microentrepreneurs in Rwanda about the last ten calls made or received on their mobile phones. This was easy to do—we just asked them to open their call logs on their handsets. In the majority of cases, the proportion of social calls outpaced business calls. This population of microentrepreneurs was the archetypal "instrumental" (purpose driven) group, according to ICT4D narratives, and yet SMS messages for the festive season, calls to girlfriends and spouses, or just little chats to say "hello" outpaced calls to suppliers and customers. In some cases, calls pulled double-duty and were for both business and leisure. I would expect to find a similar pattern of app-based downloads and informational behaviors among smartphone-using microentrepreneurs today, with time in social networks and on video sites intermingling and outpacing any business-related apps they might have downloaded. One of the core elements of the mobile device is its ability to serve our universal, noninstrumental needs—to let us connect and pass the time. We cannot exclude these things from any narrative about development.

And fourth, noninstrumentality is important on its own. Games, expression, entertainment, and leisure are core functions on mobile devices. We learn from them,[11] craft identities with them,[12] take pleasure from them,[13] and are human with them.[14] We have to be careful with our inquiries regarding productivity and production because in many ways there is nothing "better" about a spreadsheet than about playing a game of Scrabble with friends.

For these four reasons, there is no optimal level of informational production that we should expect or hope to find on the mobile device. This is especially the case if we cannot observe what kinds of things people are doing on other devices in their digital repertoires. To echo Dorothea Kleine's quote from chapter 8, if we see ICTs as "technologies of choice," which play "a vital crosscutting role in assisting people to lead the kinds of lives *they* value,"[15] those choices might or might not require engaging in informational interactions with remote services. It is up to users to find their own paths.

There is another distinct, important problem with a frame that mobile devices are relatively bad at supporting productivity and generativity—and that is that, with every passing day, they are less bad. In fact, high-end devices can be quite good, for certain tasks. One can write a long email, or even a book, on a feature phone or smartphone.[16] In Japan, *ketai shôsetsu* (cell phone novels) with short 100–200 word chapters, created and consumed on mobile devices, are quite popular.[17] In the UK, the mobile story app Wattpad is popular among teenage girls.[18] The same is true for video. For example, Max Schleser reviews the emerging "mobile pixel aesthetic," which he says echoes "avant-garde artists in the twenties [1920s]."[19] Growing numbers of filmmakers and artists are using mobiles to create experimental films and art installations at the bleeding edge of culture and technology. Schlesler illustrates just how far mobile production has come, listing commercial advertisements and music videos shot and edited on mobile devices. For these reasons, it is not far-fetched for Apple to boast of the production capabilities of its iPad in advertisements,[20] or for Microsoft to release versions of its Office productivity suite for almost every mobile device available.[21]

For certain tasks, like cleaning up photos (with filters), editing short films for uploading to social media sites (with templates), or blasting out a quick email (with predictive text), mobile handsets may already be faster and easier than PC alternatives. With bigger screens than phones, tablets begin to close the gap with personal computers for everyday writing. Indeed, with or without add-on mechanical keyboards, they almost fit the bill.[22] The digital repertoires of many people will include advanced skills and literacies to get more out of the productive capabilities of their mobile devices. Some people are amazing mobile typists, or mobile photographers. Regardless of one's skills, however, the essence of many apps is how task supportive they are—how simple an app makes it to do one thing.[23]

Yet for all of these reasons, those remaining lags and pauses that we *still* experience as people with multidevice repertoires are all the more instructive. Why do so many of us *still* wait to write a document, or to edit a long video? To make an animated cartoon? Or prepare a CV? Or run a statistical model? Or create computer code? Why do the teenagers in Cape Town wait in line to do their homework at the library instead of on their mobile devices? The answer: because the same elements and properties that helped drive the boom in mobile device adoption—inexpensive devices, usage-based pricing, personal/portable form factors, and task-supportive interfaces—tend to discourage (or at least, not actively encourage) certain forms of informational production relative to PCs.

The rest of this chapter explores two ways in which these persistent restrictions on certain kinds of informational production intersect with ICT4D and the challenge of digital inclusion: for conciseness, I compress these into "earning a living" and "being heard." Both are activities that can involve digital tools for content creation, but do not end there. Rather, they take place in global systems of power and exchange. There are political economies in play, and digital repertoires play a part in both reflecting and reinforcing what Jen Schradie calls the "digital production gap" and the maintenance of an "elite Internet-in-Practice"[24] with respect to the creation of financial and nonfinancial (cultural, social) capital.

Of course, Schradie's work—and that of many others within the effective use frame summarized in chapter 8—suggests that this production challenge probably is not reducible simply to differing access to different tools; it is, rather, a more complex matter of skills, practice, habits, and socioeconomic status. That said, *some specific scenarios*, particularly the intensive creation and manipulation of digital information, content, and indeed meaning (documents, photos, videos, and programs themselves) remain *easier* to perform on a device with a keyboard, a larger screen, an unmetered network connection, and a mouse or pointing device. To put a finer point on it, some activities are easier to perform with a *repertoire* that includes these affordances (and the skills to use them) than one that does not.

Earning a Living

One significant concentration within the practice of socioeconomic development is the challenge of improving and enabling livelihoods—helping people earn money and make a living.[25] Of course, the process of economic development cannot be reduced exclusively to a frame about livelihoods (let alone markets), and not everyone needs a job or a livelihood. That said, it is hard to deny that "development" advances when more people have access to better opportunities for rewarding livelihoods. As such, ICT4D practitioners are often interested in exploring ways that digital technologies can improve livelihoods, for example, by making enterprises more productive, assisting people with job searches, matching employers and employees, and helping people acquire the skills they need to pursue better livelihoods.

By allowing the coordination and exchange of information at a distance, creating connections to new knowledge, and allowing advertising and price discovery, mobile devices help small enterprises be more productive, and thus also help with household and individual incomes (livelihoods).[26] However, my concern here is not coordination or communication across the full range of possible vocations, but rather with the prospects presented by mobile technologies for engaging in specific kinds of informational creation and production that translate into earnings or value creation for enterprises and organizations.

In those domains, the story is different. We must grant that an aspiring director can make a video on a smartphone, that an engineer can sketch an idea for a new automotive part on a tablet in his garage, and that a microentrepreneur can keep track of her inventory using an app. These are incredible developments. Yet Disney does not create its animated movies on smartphones. Tata (the Indian car maker) does not design new automotive parts on smartphones. Carrefour (the French supermarket) does not run its inventory management and logistics system on smartphones. Nor does Samsung (the Korean manufacturing giant) write new versions of its smartphone software on smartphones—not even their own smartphones. If they are among the growing ranks of those with mobile-only digital repertoires, the aspiring director, backyard engineer, and microentrepreneur will have mobile devices, small screens, and likely metered connections, whereas Disney, Tata, Carrefour, and Samsung all have PCs, dedicated server farms,

IT support, and copious bandwidth—and a full array of mobile devices to complement all the other technologies in the mix.

My point here is that ICT4D has to be careful not to propagate a "development narrative" that treats these two general flavors of informational production as if they were likely to create equivalent value. The problem is not that it is difficult to produce content on mobile devices (one can), nor even that it is difficult to earn money for microwork performed on the phone (one can, as I described in chapter 6). Rather, the persistent issue in the shift to a more mobile Internet is that it remains relatively difficult to produce content for which one will be paid *relatively well*. It is obvious, but needs to be said: in the marketplace for professional content and informational work, producers (organizations and individuals) with mobile-only repertoires must compete with other producers who have more extensive digital repertoires.

One way we can assess how differences in digital repertoires influence effective use is to understand their contribution to the accrual of forms of capital—financial, human, social, and cultural. Rice and Pearce[27] build on van Deursen and van Djik[28] and Zillien and Hargittai[29] to suggest that some online activities are more "capital-enhancing" than others,[30] and that mobile devices are less supportive of those activities than PCs.

Andrew Maunder and I wrote about his startup Kuza.com, which allows microentrepreneurs to create web pages for their businesses, right from the web, from smartphones and for feature phones. Part of the value offered to the user was support and tips about what made a good website . . . except on the feature phones: "The inclusion of feature phones demanded degradations in the mobile experience that Kuza.com was able to deliver. These trade-offs are faced by any mobile service that wants to be accessible beyond those with smartphones . . . but it is particularly interesting to note how building for feature phones sometimes creates a poor experience for precisely the population that might need the most handholding and that could benefit most from a more evocative and helpful user interface."[31] In general, when linking ICTs to livelihoods, researchers and practitioners in ICT4D should be cautious and explicit about the mechanisms through which access translates to informational production and therefore to economic growth, or to a progressive and fair distribution of incomes in a growing economy. It is possible that some paths (trading, reporting, microwork) will indeed become more available thanks to the spread of a more mobile Internet while others (writing, editing, analyzing, curating, designing) will stay restricted to those with repertoires (devices, networks, services, and literacies) more amenable to computation and informational production. As more people are connected more of the time, the blueprint of the economy *will* shift. Yet while some doors to earning a living with ICTs may swing wide open, others will remain shut. How much money and opportunity for capital accrual will remain behind these restricted doors?

Being Heard

A distinct thread in development theory and practice is the challenge of inclusion and participation—of citizens having a stake in the trajectory and government of their societies and communities.[32] Development is not just exclusively about participation in public life, either, and not everyone needs to participate, but development is more effective when many people do so. That said, the new forms of digital production enabled by ICTs are particularly compelling within this discourse on participation, so other scholars and practitioners have linked ICTs to participation under banners like "Development 2.0"[33] and "ICT4D 2.0."[34]

Yet how many stories, narratives, and perspectives are not heard because of production barriers still associated with mobile-only digital repertoires? Some in the United States have raised concerns about the gap between "digital consumers" and "digital contributors."[35] The increasingly varied configuration of digital repertoires presents a new wrinkle in the existing research on the "production gap"[36]—adding differences in hardware and software to the demographic and skills distinctions already separating what van Deursen and van Dijk call "consumptive" and "contributory" forms of Internet use.[37]

As I mentioned earlier, writing is of particular concern. Instant messages, text messages, blog or Facebook posts, and even short emails are as easy to write on a smartphone as on a tablet or a PC. Yet the gap has not closed entirely. For composition, editing, translation, and preprint formatting, full-featured PCs are still preferable,[38] not just to data-enabled feature phones[39] but even to powerful smartphones and tablets.

However, the issue is not just the form factor or the interface. There are elements of where (in an online sense) writing occurs, where it is hosted, and how it is indexed that may be reinforcing asymmetries of production and inclusion. Marion Walton's work with various colleagues on these issues is quite instructive. In 2009, she and I looked at the early days of mobile-mediated participation in South Africa's elections, contrasting the kind of mediated participatory spaces available to PC users with the more restricted spaces available via Mxit, the most popular mobile instant messaging app in South Africa at the time.[40] Mxit users' content was not archived or searchable in the way boyd and Ellison[41] suggest is desirable for public social networks. Walton and Leukes returned to the topic a few years later, exploring patterns of political participation on Mxit and Facebook by feature phone users. They observed vigorous political debates happening in real-time across mobile social networks including Mxit and heterogeneous social networks like Twitter and Facebook. Yet they argue that even the simple act of link sharing remains challenging on a feature phone, and remind us that "user generated content on Mxit, which is used by millions of young South Africans, is unlinked, and goes unindexed by mainstream search engines and is seldom noticed or investigated by mainstream media, which rely on Google searches and 'sentiment analysis' of social media."[42]

Even on Facebook, they observed how feature phone versions of Facebook made linking difficult; what few political links were present came from smartphones rather than feature phones. Again, the takeaway is not that participation is completely absent, but rather that the terms and depth of that participation is moderated by the intersection of interface, network, and service-level architectures, as well as by users' digital literacies. I would suggest that readers do not fixate on whether a given model of mobile device affords the ability to do any specific task. Focus instead on how people with a broader range of more powerful devices spend more time doing tasks with digital configurations that are optimized for those tasks than do people who run most or all of their digital interactions through a single low-cost device. These interface compromises and constraints will not bother all users, but should be of concern to those in the ICT4D and information society communities, lest we jump to the conclusion that thanks to mobiles, the digital public sphere[43] is now equally open to all.

Support Production Scenarios

In this section, I will focus on three distinct, yet complementary approaches to address the persistent production gaps[44] associated with mobile-only digital repertoires. It is worth stressing the idea of *constrained* digital repertoires again. Keep the feature phone or the low-end smartphone in mind, not the flagship devices. You will also notice that in this section I am writing about repertoires, rather than specifically about mobiles. It is the best and most optimistic frame I can muster, and many of the solutions will require augmenting *repertoires* (through adding devices or skills), rather than asking the mobile handset to be something it is not very good at being.

Improve Production on Low-End Mobile Devices

The production capabilities of low-end devices in the hands of low-resource users are getting better over time. Smartphones will replace feature phones for many resource-constrained users, and smartphones themselves will continue to improve. The availability of phablets, better predictive text on keyboards,[45] and better software and app experiences supporting productivity are all likely to continue to narrow the hardware and software dimensions of the production gap without special intervention or pressure from the ICT4D and development communities. That said, some additional areas may be of particular concern to ICT4D practice.

Replacing Text with Speech

Chapter 8 described how low literacy and numeracy present challenges for broad-based, effective use of the Internet in general.[46] To remove technical barriers to mobile informational production among low-resource communities, further efforts and interventions should target speech as an alternative to text. After all, every mobile phone has a speaker and a microphone, no matter how cramped its screen and buttons might be. As I mentioned in chapter 2, researchers are working on extending and refining the use of icons, audio, and video prompts, and other shortcuts to make manipulation of the mobile phone easier for low-literacy users.[47] There is still a great deal to work out about navigating the web via voice. For example, how to toggle from voice directions to units a computer would understand;[48] how to handle file storage and search of voice data by and for oral cultures;[49] and how low-literacy users can make sense of voice-based interactive systems.[50]

IVRs extend some indirect Internet access to basic phones, and already effectively support programs in agriculture[51] and health.[52] However, the specific question here is about production. CGNet Swara, described in chapter 3, hints at new forms of content creation via voice. Better speech interfaces could help reduce (although perhaps not eliminate) the privileged structural advantage text (and textual literacy) seems to hold in information/post-industrial societies. It could fit well with the oral traditions of many other communities around the world, and could, if coaxed and nurtured, lead to a greater diversity of voices and perspectives having a home on the Internet.[53]

The key to these interfaces may be beyond interactive voice response (IVR), in the realm of full-blown speech recognition, the "spoken web,"[54] and "natural language processing."[55] In these cases, users can use their own words, in their own language, to make their computational and informational needs understood by computers, and whole sentences and paragraphs can be transcribed and possibly translated by machines in the cloud. Speech is an exciting domain in mobile computing in general, as evidenced by the branding and promotion of the speech recognition services on each of the major smartphone platforms—Apple's Siri, Android's Google Now, Microsoft's Cortana.

Yet so far, these services are not universally accessible. This is for two reasons. First, natural language processing is computationally challenging and requires large files of memory, so it happens in the cloud, not strictly on the device. Thus tariffs might apply.[56] Second, the process of building training materials and configuring speech recognition is costly. That means less-prosperous users in small markets are still waiting.[57] High-quality speech recognition in a wide variety of languages is both more challenging and potentially more revolutionary, since the diversity of content, services, and interfaces available in the world's most commonly spoken languages, like English, Chinese, German, French, Russian, Japanese, Portuguese, and Spanish, dwarfs that available to speakers of hundreds of other languages.[58] User-generated content, especially on social networking services, may begin to erode these differences, and automated and crowdsourced community translation will help, but there is a long way to go before the Internet—mobile or not—is as useful to someone who speaks a language with ten thousand speakers than it is to someone who shares a language with 300 million speakers.[59] In order to support voice-intensive production scenarios in the developing world, any rollout in mobile cloud computing will also require simultaneous advances in the costs and reliability of connectivity.[60]

Training to Improve Mobile-Centric Digital Literacies

A completely different approach to closing production gaps is decidedly nontechnical. Interventions around promoting *mobile* digital literacy may help those with mobile-only and mobile-centric digital repertoires. As I mentioned in chapter 8, there is a large research literature and community of practice around understanding and promoting digital literacy that predates the mobile Internet, and is bound up with an accumulating body of evidence about digital inequality and persistent differences in digital skills.[61]

On the one hand, some have argued that mobiles demand (and encourage) distinct digital literacies from PCs,[62] and thus digital literacies and learned practices in digital repertoires may not move seamlessly between devices. Examples of these mobile-specific literacies include: the skillful integration of multiple languages or shorthand "txtspeak" into text messages; "code switching" in ways that exploit and transform linguistic forms;[63] image-making and refinement on small screens;[64] the "prepay literacies" around cost discussed in chapter 9;[65] the use of specifically mobile affordances (like voice and location-based services) in information searching;[66] and the familiarity with concepts from basic mobile interfaces rather than PC-based Internet as the baseline for navigation of the Internet on mobile devices.[67]

Marion Walton, my coauthor on the telecenter project described in chapter 8, has done extensive work exploring mobile literacies and training programs in the South African context. She and her colleagues have created interactive reading experiences specifically configured around consuming and authoring content on mobile devices. For example, they created a serialized teen function novel called Kontax, accessible via a mobile website, that invites its readers to "express themselves" with stories or posts of their own. Their work raises important and unsolved tensions between the formal literacies, styles, and formats associated with textual production on PCs and the informal, practical, everyday, intimate literacies learned through regular mobile use, such as textisms, abbreviations, and sparse punctuation.[68] Similarly, Watkins et al. describe piloting training programs around smartphones for "participatory content creation" for a citizen journalism program in India; however, they signal how the formal journalism institution behind the pilot was uncomfortable with the breaking down of the hierarchies between producer and reader.[69] These are early examples of what should be a burgeoning set of initiatives to integrate mobile digital literacies via training and software alike.

On the other hand, it is probably unwise to conceptually separate mobile literacies from other digital literacies, since the functionalities of mobile telephones and Internet devices continue to converge, both on the devices and in the cloud. Instead, digital literacies support individuals as they manipulate and create data, information, and meaning residing on distant, unseen, largely mysterious servers via whatever device happens to be available.

To the extent that any school or institution has the resources to teach digital literacy to its low-resourced, mobile-centric or mobile-only constituents, it would be important to try to update definitions and practices of digital literacy to include production on mobile devices, though not at the expense of combinatory practices, in the spirit of Madianou and Miller's polymedia[70] and Jenkins's transmedia.[71] This may pay particular dividends when working with NGOs and other "communication for development" practitioners already at work around the world. Those with the motivations and incentives to construct change campaigns may need assistance in updating methodologies and approaches to take better advantage of mobile Internet modes.[72] For example, Mozilla, the makers of Firefox, have a remarkable "Web Literacy Map" sketching "the skills and competencies needed for reading, writing and participating on the web."[73] The overview document, as of early 2015, did not mention the word "mobile" even once. It probably should.

Convert Mobile-Only to Mobile-Centric Repertoires

Better affordances and better training would help close the production gaps between mobile and fixed modes of Internet use. However, each of these general approaches may be difficult to support at a large scale, for billions of would-be users. It is too optimistic to suggest that the answer to the production gap lies with making mobile phones converge toward PC functionality, or even in training everyone to get the most out of their mobile devices. Mobile devices are so compelling for consumption and coordination scenarios, and as complements to other hardware form factors more amendable to production, that it seems counterproductive—not to mention unattractive to the marketplace—to turn a mobile handset back into a production-optimized PC.

Further attempts at converged devices will appear. The open-source software company Canonical has proposed doing just this for midrange ($300–$400) mobile devices, using its Linux-based Ubuntu OS to power a handheld that could be expanded via a keyboard, mouse, and external monitor,[74] but only time will tell if this will be successful at the critical sub-$100 level. Microsoft's 2015 version of Windows will further close the gap between phone and PC operating systems, allowing apps to run on both platforms and carrying a common look and feel,[75] but a "phone" is more than the operating system, and even a mobile device that runs Microsoft's full-power Office productivity suite is a better complement to a PC experience than a replacement for it, at least for certain information production scenarios. Similarly, the hybrid space between the phone and the tablet, although easier to bridge physically than the one between phone and PC, does not seem to be the logical end-point either. Phablets may be fantastic for media consumption, but due to issues including battery life, cost, and form factor, the phablet that is large enough to facilitate easy production may not be small enough (or cheap enough) to fit with all users' lifestyles.

Stay Enthusiastic about Low-Cost Computing Options

Instead and in the meantime, we can step back from the exclusive focus on the mobile handset. The discussions of digital repertoires and the After Access Lens remind us that making *other* form factors more affordable and accessible is also important. Specifically, frugal engineering efforts to reduce the cost barriers to families and small organizations purchasing low-cost computing devices like tablets and PCs should also continue. If a base-of-the-pyramid household can afford mobiles for each member and, perhaps, a tablet to share, that would create more options in otherwise mobile-only households, and would more closely approximate the specialized digital repertoires enjoyed by more prosperous families in the Global North. As the hype (and yet-unfulfilled hope) around the "$35" ultra-low-cost Akash Tablet in India suggests,[76] there is a battle heating up to serve consumer demand and get the costs of seven-inch tablets down below $100.[77] Extremely inexpensive PCs, like the tiny, kit-like Raspberry PI,[78] or low-power units running Android,[79] are also chipping away at what people have long considered the low-end of what a computer costs.

There is much to like about low-cost tablets. Big screens, in particular, help ease composition and image editing. The addition of external keyboards and pointing devices makes the tablet even better. There are myriad organizational settings, some private (like families[80] and small enterprises), some more public (like schools[81] and large organizations), where tablets can help expand the production/contribution/processing capabilities of the people in those groups.

In this vein, I am predisposed to look fondly on the relatively famous, relatively controversial One Laptop per Child (OLPC) initiative.[82] Launched in 2006, OLPC focused on educational scenarios, in the first iteration, via kid-friendly laptops, and later by rugged tablets running a variant of Android. In all cases, the OLPC model stressed constructivist learning and represented a faith in the power of technical and pedagogical innovation to overcome resource constraint, poor schools, and decades of underinvestment in education in many developing countries. Ministries of Education in countries including Rwanda, Bolivia, and others bought OLPCs by the truckload. Critics were concerned about the rosy scenarios,[83] rather light pedagogical models,[84] and general "utopian" approaches[85] associated with the OLPC initiative. Yet viewed through the After Access Lens, OLPC hits many of the right marks, even with a vision originating from 2006. With creative mesh networking[86] and laptop and tablet form factors, optimized for challenging environments, the hardware stressed coding and creativity under conditions of network constraint. So did Sugar, the software underpinning the collaborative tools. OLPCs cannot solve all the challenges that educators and students face in developing countries. However, refinements to pedagogy, continued experiments,[87] and launches of similar initiatives should leave observers cautiously optimistic about the longer-term prospects for educational technologies in low-resource settings.

Continue to Support Public Access

The other way to encourage and enable users to shift from mobile-only toward mobile-centric digital repertoires is for policymakers, governments, civil society, and entrepreneurs to continue to support public access venues as complements to private mobile Internet use. Horner cautions: "whilst the mobile internet may help to overcome connectivity gaps in the short term, focus on the mobile internet must not undermine efforts to achieve universal access to desktop computers and advanced fixed-line fibre optic networks. Mobile, computer, and fixed line access must be thought of as mutually compatible as synergistic technologies, to which all should have access."[88]

In the telecenters study I described in chapter 9, Marion Walton and I found public access venues provide many nonsubstitutable benefits to visitors who have Internet-enabled phones: keyboards, big screens, safe and quiet places to work, printers, training, and of course, subsidized bytes of free access. All these can help people who own only a mobile phone to get all the value of a PC, when they need it.[89]

It is important to stress that access to shared public computing was nevertheless *scarce and occasional* for the teens in our study. In the libraries, students were limited to thirty-minute slots (via a sign-up sheet); their use was monitored, and noninstrumental uses like video games, Facebook, and YouTube were discouraged. Once walk-times and waiting-times were factored in, access to the Internet was neither free nor unfettered. Yet some access to PCs was preferable to none, because for some production scenarios, like doing research for homework assignments, or looking for jobs (preparing CVs and sending emails), students preferred and were likely more effective users of the PCs available to them.

Maintaining and even expanding this availability of public computing resources will not be easy, however, as the narrative around public access may be shifting. If decision makers act as if mobiles and PCs are perfect substitutes, rather than as complements in digital repertoires, support for shared access venues will be increasingly under threat. Indeed, many scenarios no longer require a trip to the telecenter or cybercafé. From booking bus tickets to posting social network status updates, to playing games and looking up treatments for a baby's rash, the mobile Internet connections in peoples' pockets will allow them to cut back on public computing, or never start at all. This shrinking demand may make the prospects of running a "sustainable" or even profitable public access venue—not easy to begin with—even more difficult to achieve, since venues cannot count on loads of paying customers for these features.

Nevertheless, there might be a lasting role for telecenters and other public access venues, even in cases where many library or telecenter users might own a data-enabled mobile phone.[90] In addition, community access points may become community hotspots in the style more common in the Global

North.[91] Even if states are not willing to set tariffs for mobile data, or to subsidize mobile-data tariffs for resource-constrained users, they may be willing to continue to provide free or subsidized bandwidth via Wi-Fi in community libraries, schools, telecenters, and other public-access venues. Revisiting universal service obligations and other sources of nonrevenue seeking investment may be necessary to provide primary access (if needed) to those still without mobile phones,[92] and occasional, potential, complementary access for the billions more with a phone. For as the thought exercises in this piece (reflecting on "earning a living" and "being heard") suggest, there are times and scenarios for which widespread affordable access to non-mobile computing resources is still important, particularly for digital inclusion and the development outcomes so many of us care about.

Conclusion

This chapter introduced two themes of interest to ICT4D—earning a living and being heard—that require "informational production scenarios." I argued that the configuration of digital repertoires plays some role in the likely success of these activities, and that people with mobile-only digital repertoires face additional challenges in converting their ideas (and clicks) into revenue or influence. The second part of the chapter highlighted three distinct ways researchers, practitioners, and technologists could minimize the extent to which mobile-only digital repertoires present barriers to production and inclusion.

In a sense, the most important takeaway from the chapter is about framing what mobiles (and a more mobile Internet) have meant for the "digital divide." I echo other researchers[93] in stressing that practitioners, theoreticians, and policymakers should be wary of proclaiming the closure of the so-called digital divide, due to mobile phones alone. We should carefully avoid claims that equate two repertoires—one with a $40 feature phone and the second with a smartphone and a PC and whatever array of digital devices a prosperous person might have. Both repertoires allow access and use of the Internet, but not necessarily the same Internet, not the same uses. Production is one place where the mobile-only and mobile-centric digital repertoires are particularly and critically divergent.

This is not to say that production does not occur via mobile devices. Yet when we assess the spread of informational production via mobile devices we should not let the (absent) perfect be the enemy of the (nearly ubiquitous) good. For example, Naomi Baron's review of reading on phones suggests that while mobile devices may not be as good for in-depth, "deep," or repeated reading as printed pages are,[94] they may bring more availability than printed pages. Similarly, video, or coding, or photography may not be "as good" when created on mobiles as on professional equipment, but the extended reach is phenomenal. Yet these lingering shortcomings, concentrated as they are around information production, raise tensions in the "development narrative" of the meaning of inclusion in the information society. We should remain aware of persistent, replicated stratification of the mode of production in digital goods, digital value, and digital meaning.

Let me close the chapter with a bit of admittedly glib speculation. Sometimes I wonder whether any of the recent articles in blogs and the popular press about the "death of the PC"[95] or the "post-PC era"[96] were written on anything other than a PC. (This book, too, was written on a PC.) In the same spirit, I would guess that just like me, readers of this book are still unlikely to be willing to give up entirely their PCs for a mobile-only lifestyle. A mobile-mostly (mobile-centric) lifestyle is quite appealing, as long as there is the security of a PC around from time to time to help with some occasional—but valuable—heavy lifting.

11 Circumscribed Structural Roles

Jenna Burrell notes, "Effective use is not just about skills and literacies, but also about being empowered to participate in the production of the internet."[1] How invested, engaged, and influential are those with mobile-only digital repertoires in shaping the structure of the Internet itself?

Chapter 9 described the metered mindset, a constraint that affects almost all mobile-only users. Chapter 10 sketched some persistent limits to certain forms of information production, constraints that affect only some users directly, but all of society by extension if some voices are not heard and some productive activities not rewarded. This chapter will explore the third category of constraint in the After Access Lens: the fact that several of the same properties that have made mobile devices so popular, powerful, and easy to use may also discourage users from influencing the trajectory of the Internet itself. Harmeet Sawhney uses the term "the character of the Internet" to frame the changes underway in a shift toward mobile.[2] I'll echo that frame in this chapter, suggesting that the shift to a more mobile Internet alters the terms on which individuals can participate, in coding, user agency and control, content and software distribution, and editorial gatekeeping. As was the case with information production, these constraints may be felt *directly* only by a subset of would-be coders and entrepreneurs, but in aggregate they have implications for the Internet we all share.

This topic is vast, so for the sake of parsimony I will frame several mobile-specific critiques as four broad, overlapping departures from the initial design ideal (e.g., character) of the World Wide Web. That ideal, articulated decades ago by some of its inventors at CERN as "*a pool of knowledge that is as easy to update as it is to read,*"[3] has been remarkably durable and flexible. However, that ideal (if not the network) is strained by increasingly varied uses of the Internet Protocol (IP), connecting everything from streaming video to home thermostats on the evolving "Internet of Things."[4] IP enables services and systems that have nothing to do with the World Wide Web of hyperlinks, browsers, or expanding webs of interconnected pages of content as envisioned by CERN. Whether this increased diversity is "good" or "bad" is well beyond the scope of the chapter and the book—after all, the web is just one way to use IP. However, the ways in which the shift to a more mobile Internet may be accelerating this departure away from the web ideal—and in so doing, altering the relationship between mobile-only users and the Internet itself—is something this chapter can address. Of special concern are the implications of these shifting relationships for ICT4D.

Four Structural Circumscriptions

In this section, I introduce four general circumscriptions—ways in which inexpensive devices, usage-based pricing, and, particularly, task-supportive designs have combined to attenuate power imbalances between individual mobile-only users and other central actors. There have always been system administrators, editors, gatekeepers, investors, power users, and other structurally powerful actors on the Internet, but the shift to a more mobile Internet, albeit one with more users than ever before, further exacerbates the differences between users and producers, customers and gatekeepers. The first three circumscriptions involve the experiences of individual users. The fourth pertains specifically to coders, hackers, and makers.

It Is Hard to See the Whole Internet (from a Phone)
In 2010, Chris Anderson debated Michael Wolff in *Wired* magazine in a feature titled "The Web Is Dead, Long Live the Internet." Said Anderson:

Over the past few years, one of the most important shifts in the digital world has been the move from the wide-open Web to semiclosed platforms that use the Internet for transport but not the browser for display. It's driven primarily by the rise of the iPhone model of mobile computing, and it's a world Google can't crawl, one where HTML doesn't rule. And it's the world that consumers are increasingly choosing, not because they're rejecting the idea of the Web but because these dedicated platforms often just work better or fit better into their lives (the screen comes to them, they don't have to go to the screen). The fact that it's easier for companies to make money on these platforms only cements the trend. Producers and consumers agree: The Web is not the culmination of the digital revolution.[5]

To put this in terms of the elements of the After Access Lens, the fact that "Google can't crawl" (discover) so much of the mobile-generated content is a constraint associated with *task-supportive apps*, and means the return of "walled gardens"—apps and services where data is either nonaccessible to outsiders, or simply invisible and never archived.[6] The result is the first structural circumspection—an Internet where it is not as easy to read the pool of human knowledge as it might have been when the web logic was ascendant. Mobile-only users here are not at a particular disadvantage when it comes to search and discovery, but rather are helping to structure the "character" of the Internet in this way by downloading and using these apps that "just work." Billions of microdecisions, each day, by users around the world, are eschewing the web/browser logic for the convenience of mobile apps. Tech blogger Danny Crichton elaborates: "From the user's perspective, compiled apps are easier to discover, seem more natural, and perform better. . . . It's truly a sad moment, given that we are sacrificing so much of the Web's best qualities for proprietary native apps. There is no way to construct URLs to apps, nor any method to hyperlink to specific content within an app container, a concept called "deep linking."[7]

For ICT4D, the sheer explosion in the amount and variety of content and services happening thanks to advances in mobile Internet systems is a great thing, full of promise and potential for digital inclusion. Yet a byproduct of these choices—the unintentional fragmentation and balkanization of that content and services, and their lack of permanence, accessibility, discoverability, and interconnectedness—should be of concern. Locking up knowledge or know-how in an app, no matter how good that app may be, comes with tradeoffs about how that knowledge (or code) may be repurposed and reused in serendipitous ways.

It Is Hard to Control Content (from a Phone)

Another circumscription involves the control of content, in what Sawhney describes as the long-term battle between bottom-up "chaos" on the conventional Internet and top-down, centralized "order" of its emerging mobile forms.[8] One example of the tensions over control is digital rights management (DRM): restrictions around rights to copy, transmit, or alter content purchased or rented from content providers and digital intermediaries. Internet commentator Cory Doctorow elevates his concern over digital rights management to one of his "laws" of the web: "anytime someone puts a lock on something that belongs to you and won't give you the key, that lock isn't there for your benefit."[9] Not everything about the mobile is closed, nor is everything on the general Internet open. However, to echo Crichton's remarks above, the shift to an ecosystem dominated by apps (and time spent in apps) has created a new array of locked and closed content and service relationships between users and vendors, between consumers and providers, rather than between co-creators and peers. Unpacking the rhetoric of a Nokia report on new mobile media features, Jonathan Little said this almost a decade ago, even before the rise of the app stores:

In the mobile world, citizenship is predicated on a traditional liberal concern with an individual's participation in the marketplace. DRM is presented as a necessary tool for regulating the consumer's ability to use a media product/cultural text. Limited cultural and social rights translate into limited, regulated, and controlled communicative rights. From a social democratic perspective, restraints on an individual's ability and right to communicate are to be abhorred. But Nokia's DRM White Paper deploys the rhetoric and logic of the marketplace to circumnavigate democratic rights. From this perspective, images and sounds are not culture and language. They are commodities. Thus, corporations should be allowed to regulate the conditions of purchase and use.[10]

DRM and other restrictions on apps and content matter for ICT4D. For example, apps may not cross international borders with their users;[11] laws covering apps use may differ, and those in the market for secondhand handsets may find that many of the apps are tied to identities that did not transfer with the phone.[12] Lots of people (and phones) shift across international borders. For all the emphasis on user agency and empowerment, the specifics of the end user license agreements (EULA) and DRM protocols can make the use of many mobile devices more tightly bounded by the OS, app, and hardware providers than had been the case with PCs and the conventional Internet.

Indeed, and ironically, these power imbalances also influence not only what a user might purchase, but also what he or she might contribute or create. As "platforms"[13] like Facebook and Twitter replace standalone websites and blogs as the supposedly public venues for the production and exchange of human knowledge (to echo the CERN framing of the goals of the World Wide Web noted earlier), users may be licensing or transferring ownership or control of their own content to those hosting those platforms. John Battelle, an expert on search, details some of the implications for businesses of relying too heavily on any of the platforms as their interface with customers and audiences, reminding readers that "'full, top to bottom control' means a lot more than just the chrome finishes on your website. It means controlling all the data created by interactions on that site, including if and how you share that data with your consumers and your partners."[14] Battelle is writing for companies in the Global North, but his words might soon apply for all sorts of organizations and institutions in the Global South. Users or organizations trying to maintain control of their content using only their phones (as opposed to hosting their own websites and servers, for example) are going to face an uphill climb. Feature phones are no help here. Moreover, though it is not impossible to navigate and contribute new content to an open web via smartphones, it takes extra work to do so when the platform-based alternatives are so compelling; when they "just work."

For ICT4D, the issues of content and control may complicate efforts to use and encourage mobile Internet platforms for community participation. For those who suggest that open systems should be at the core of ICT4D efforts,[15] the concentration of power on platforms, in app and content stores, in the popular mobile operating systems, and on a handful of social networking services and popular publishing platforms, should be of significant concern.

It Is Hard to Stay Private (on a Phone)

A third structural circumscription concerns security, and the related issue of privacy. The same properties that make the device so compelling (personal intimacy, wirelessness, task supportiveness) can exacerbate threats to privacy. As Edward Snowden's revelations about the U.S. government's electronic surveillance programs in 2013 made clear, other eyes may be privy to many mediated interactions occurring via mobile devices.

This is not to say that spying and snooping is limited to mobile devices (or to the U.S. government[16]), but mobile users are particularly ill equipped to prevent it; devices seem to leak personal data and whereabouts as part of the price of their use.[17] For example, during the pro-democracy protests in Hong Kong in 2014, protesters embraced FireChat, a peer-to-peer (P2P) proximity-based messaging platform originally designed for festivals. Many FireChat users thought they were more "secure" since they communicated via pseudonyms only, but a FireChat spokesperson stressed that they were no more secure or private than Twitter or Instagram.[18] Meanwhile, in Egypt, authorities used the location features in the Grindr dating app to enforce anti-homosexuality laws.[19] China has been willing to crack down on Internet content, not just at its borders through its firewall,[20] but also internally on SNS sites like WeChat.[21]

The implications for ICT4D here are clear. At a 2014 conference I attended on "Mobile for Development," the room's conversation turned to the uneasy tension concerning the reality that some of the same elements afforded by mobile Internet devices—legibility, enumeration, tracking, targeting, and standardization—that could be used to support state efforts to promote development and deliver services via the mobile channel could also be used for surveillance and control. User goals like privacy, autonomy, agency, and freedom are at risk when billions of people willingly carry "little brothers" (a play on Orwell's "Big Brother") everywhere they go.[22]

It Is Hard to Change the Internet (from a Phone, or for a Phone)

The final structural circumscripton I will discuss involves the tinkers, hackers, coders, makers, and entrepreneurs around the world, and the degree to which the shift to mobile means that someone else (a centralized authority like a platform, an algorithm, or a regulation) is more likely to control what those creators can share and accomplish. These structural restrictions on changing the Internet itself involve a particularly specialized and abstract set of constraints. Yet the goal here is to identify barriers that may stand in the way of those who otherwise would wish to do so, with the assumption that digital inclusion demands that those barriers are minimized.

In the influential book *The Future of the Internet—and How to Stop It*, Jonathan Zittrain focuses on the idea of generativity; how, through leverage (power), adaptability, ease of mastery, accessibility, and transferability, technologies have the capacity to "welcome contributions from outsiders as well as insiders."[23] Marking the launch of the original iPhone in 2007, Zittrain suggested:

The PC revolution was launched with PCs that invited innovation by others. So too with the Internet. Both were generative: they were designed to accept any contribution that followed a basic set of rules (either coded for a particular operating system, or respecting the protocols of the Internet). Both overwhelmed their respective proprietary, non-generative competitors, such as the makers of stand-alone word processors and proprietary online services like CompuServe and AOL. But the future unfolding right now is very different from this past. The future is not one of generative PCs attached to a generative network. It is instead one of sterile *appliances* tethered to a network of control.[24]

This network of control manifests in different ways depending on the goals of the user.

For individual users wanting to customize their devices, things are not quite as dire on this dimension as they were in 2007. Lobato and Thomas note that thanks to a dizzying array of apps, most users have a great deal of choice over what software they load, and thus how the device interacts with the Internet.[25] In addition, skilled users or third-parties can "root" or "jailbreak" a mobile device, replacing the operating system installed by the manufacturer with one with custom modifications.[26]

Individuals who want to control their network might not be so lucky. The very essence of delivering data over cellular wireless systems creates incentives for what Powell calls "technical enclosures." She argues that MNOs and mobile service providers have incentives to create propriety standards to differentiate themselves from competitors. In addition, the fact that providers have to buy spectrum in blocks has created "a return to control of the entire process of data transmission—from sender to receiver," instead of encouraging bits to traverse (as a best effort) the network of networks on the Internet as a whole.[27] The temptation to manage traffic, through shaping, throttling, and capping,[28] is not confined to mobile networks (fixed ISPs do it too),[29] but the incentives for MNOs to do it may be higher.

For people with mobile-only digital repertoires, who want to make apps, coding may be an uphill battle. It is theoretically possible to create an Android app using development software that runs Android, or a Windows Phone app while running Windows Phone, via TouchDevelop.[30] At present, to create an app for the Apple app store, one runs Xcode or the IOS_SDK, which runs on Apple's OS X desktop software. With the right skills in the repertoire, the technical barriers to phone-based coding are not insurmountable, but it is hard to code elegantly and rigorously from a phone. For the most part, people write software for mobile platforms on desktops and workstations, because, to echo the previous chapter, it is easier and more powerful to do it that way. Mobile-only users, particularly on feature phones, are not included in this part of the (re)generation of the system itself.

Thus, the most important and far-reaching parts of these constraints on generativity concern anyone who wants to code *for* the mobile platform, rather than *on* the mobile platform. This is a broad topic and a small section may not do it full justice, but I can highlight a few of the issues confronting developers, particularly in the Global South, as they try to write apps for global smartphone ecosystems.[31]

App Stores as Bottlenecks

Pon, Seppälä, and Kenney offer one helpful frame on these issues, arguing that the bottleneck has shifted from control of the operating system on the device to control of the stores and content experiences that link to the OS.[32] Apple controls its proprietary, closed iOS and the accompanying app store. However, the case for Android, by far the dominant mobile OS in the Global South,[33] is more complex. Elements of Android itself are open source, so firms like Amazon, Samsung, Xiaomi, and others are free to "fork" Android, modifying it in ways that appeal to their customers. But as Pon, Seppälä, and Kenney explain, "when the OS is commodified, firms try to differentiate and lock in users with cloud-based services, including online marketplaces, communications services, and cloud storage. Strategic development of the interfaces, or APIs, can help firms create more control and differentiation in these services."[34] Pon and his colleagues further suggest that there is a struggle for control of user activities and loyalties playing out in these complex ecosystems, in which layers of open and closed software systems (OS, store, API, app) can lie top of each other.[35] Eaton and his colleagues identify this unfolding process as a "paradoxical and ambivalent relationship between control and generativity" at work in the mobile ecosystems.[36]

App stores and APIs make it easy to build powerful software that links with other services, and make it easy to distribute the software to mobile devices when complete. The million-plus apps on iOS and Google Play are testament to the strength of the model, as are the copious returns to the platform owners as they charge 30% on each app transaction. Yet these persistent bottlenecks and points of control have particular implications for would-be coders in the Global South. Cautioning that the "'walled garden' approach could end up defining the entire Internet,"[37] Surman and his colleagues from Mozilla put it this way: "without permission from a platform provider, creators can't meaningfully get their wares onto devices—or to the users of that platform."[38] Coders from any market may find that the curators at the major app stores take issue with their designs, and decide not to allow them. Even if their app is allowed, developers must fight for visibility, ideally becoming one of the top apps in a category if they help to sell well through the stores.

Surman and his colleagues lay out Mozilla's vision for "a user driven mobile Internet"[39] with "software and an operating system that make it easy to create any content—personal apps, a web page, a video—that posts directly to the Internet and doesn't require membership in a social network or approval from [a] gatekeeper."[40] In promoting *literacy* that starts with exploration, through building and connecting, Mozilla hopes to promote more inclusive participation in the stuff of generativity.

To make this happen we are working on practical ways to make mobile content and the app economy into something as open as the web. Part of the process involves encouraging people to make apps and content using web technologies such as HTML 5, which makes it possible for content and services to run on any smartphone without going through an app store first. From there we will need to find ways to popularize the distribution of apps and services that are more open than current app stores. We're going to experiment, for instance, with sideloading, SMS distribution, email distribution, and hyper local or carrier run content delivery systems. . . . All of these things have the potential to create open distributional patterns that are more like the open environment of the desktop web and to provide room for many to build growing businesses.[41]

This is a comprehensive and optimistic vision of technical efforts that could address structural problems in the current configuration of mobile ecosystems. It is worth tracking and supporting these efforts.

Small Market, Small Producer Woes

However, there are extra challenges with the app store model facing developers in small markets. The current app marketplaces do not support payments in all countries, meaning a developer in, say, Cameroon, may have difficulty getting paid for the app she wishes to place on the Apple, Microsoft, or Google (Play) stores. In 2013, Google supported prepay app payouts in less than 40 countries,[42] and almost cut off Argentinian developers.[43]

Nor is the generativity challenge limited to the big platforms like Microsoft, Google, and Apple. The value chain for wireless services is complex and difficult for small firms to navigate.[44] The hundreds of mobile network operators around the world hardly represent a unified point of contact for negotiating rollouts of a USSD or SMS service, or to negotiate "zero-rated" airtime like Facebook and Wikipedia have managed to do. They are powerful, and they are many; scale is hard to achieve if MNOs can block your path.[45]

The result of this generativity challenge is a landscape where there is not as much diversity in content and in app ecosystems as we might want there to be. Surman and his colleagues offer an observation about India: "The top ten Internet sites in India, as measured by traffic have just two fully Indian representatives: the *India Times* newspaper and the mobile shopping site FlipKart. The rest are Indian versions of Google Properties, Facebook, Yahoo, and Wikipedia. You see similar trends across the globe, with the exception of China, which has enough critical mass to develop its own popular in-country online brands."[46] (And, I might add, a strong enough firewall to keep international competitors off the Chinese Internet.)[47]

Being a developer in a small market is not easy. Sarah Wagner and Mireia Fernández-Ardèvol studied the app development communities in Argentina and Bolivia in 2013–2014. They found differences between the former, where developers seemed able to target international business customers on IOS and Android, and the latter, where developers focused on local institutional buyers, particularly the national government and NGOs. Yet in both cases, constructing apps for the consumer markets was considered too difficult due to the challenges of discoverability. Instead, developers were building for enterprise customers or were working through intermediaries. (In Bolivia, this included more of a stated interest in apps that promoted national development goals.) They found that "in both countries, app distribution challenges and competition on the stores disconnected developers from individual users in their countries."[48] The researchers asked whether regulator interventions or monetary support would be necessary to promote app development for small local markets.[49] App development is a rough business, with low returns for many small entrepreneurs.[50] Local developers in small markets might be attuned to local needs but, in small markets, their chances of breaking even or making money on those small-market apps might be much lower.[51]

Building Local Clusters

Thus, particularly in a small market, the lone coder is not as likely to succeed as the one situated in an environment supporting her, so a quite different part of the digital livelihoods puzzle is to expand participation in the direction of mobile Internet technologies more broadly. Local *content* is important to mobile users,[52] and its cultivation has long been a priority for those who seek a more representative and universally relevant Internet. Yet in the case of mobile Internet, and apps in particular, there is also the issue of local *products and services*.[53] This is part of a great reframe in ICT4D, away from design *for* toward design *by* a greater variety of Internet stakeholders.[54] Local production and services come from local code, local coders, local products, local knowhow, local mentoring, local financing, and local enterprises.

The answer may be to look for help in the company of others. At the risk of sounding like a brochure for Silicon Valley, or New York, or Bangalore, or Cape Town, or Boston (each a region I have been lucky to live and work in), successful digital products emerge more often from dense clusters of specialized talent and practice.[55] Backyard workshops and dorm rooms are great places to start, but it does not hurt when those rooms are down the street from venture capital, decent infrastructure, and a thriving labor market for tech talent. Entrepreneurs in Nairobi, Accra, Yangon,[56] and dozens of other cities in developing countries can and will compete, but there are chicken-and-egg style challenges associated with bootstrapping digital clusters.[57]

The leaders in the global technology industry play a role in anchoring, nurturing, and seeding regional clusters. Facebook, IBM, Google, and Microsoft (my own former lab!) of course have research labs or R&D centers or both in various cities across the developing world (Nairobi, Sao Paulo, Cairo, not just in Bangalore and Beijing).[58] Yet local players are important, too: for example, a Kenyan firm has an interesting approach to delivering low-cost video on demand (in a manner that fits the Kenyan context) that uses the inexpensive kit-computer Raspberry PI as the basis for (TV) set-top boxes.[59] It may be an uphill battle to compete with low-cost manufacturing hubs in China and in Southeast Asia at scale, but companies are trying.

To match the global and the local, innovation hubs can help nurture startup companies. Nairobi's iHub is particularly well known,[60] but there are dozens more around the continent and throughout the rest of the developing world. For example, infoDev, a multidonor program affiliated with the World Bank, supports mobile hubs in fifteen countries around the world.[61] There is still work ahead to perfect the hub model for optimal impact—for example, to find the right balance between nurturing enterprises and nurturing the cluster around them, and balancing pure venture money with more patient (but perhaps less demanding) donor funds.[62] Yet, as Loren Treisman at the Indigo Trust (a foundation in the UK) put it, there is considerable hope that "a combination of philanthropic, institutional and private sector investment in technology innovation hubs, early stage tech startups and tech for social change projects will have a catalytic effect in stimulating both economic growth and improved social outcomes."[63]

"Free" Bits—Internet.org and the Complications of "Zero-rating"

In chapter 9, I described the "metered mindset"—the looming, persistent deterrent to surfing, browsing, and effective use of the Internet facing those who must pay "by the bit" or face a monthly cap on their Internet consumption. I also described how one of the most interesting responses to the metered mindset has been the emergence of "zero-rating," in which a content or service provider enters into an arrangement with a mobile network operator to subsidize the data tariffs associated with use of their service. In 2015, zero-rating stopped being simply interesting, and became controversial.

In this section, I want to return to zero-rating, treating it as the capstone to this chapter on circumscribed structural roles. The debate around the simple act of picking up the tab for bandwidth has become something of a watershed moment in the public awareness of how the shift to mobile is transforming the "character" of the Internet. As the window for me to make changes to this book closed in early May 2015, the attention of the international Internet policy community was focused on Facebook's Internet.org, and whether its zero-rating practices in India were a violation of the principles of net neutrality.[64] Events were moving quickly, and the dust had not settled on the fracas by the time this book went to print.

Facebook was not the only group using zero-rating in 2015. As I described in chapter 9, Twitter, Google, and Wikipedia were, as well. The advocacy group Digital Fuel Monitor found more than ninety zero-rated services active in the OECD in 2014.[65] Indeed, in India, at the same time that

Internet.org had partnered with the Reliance MNO, a rival MNO, Airtel, had launched its own zero-rating arrangements with media content and retail content providers.[66] Backlash and debate about both initiatives gathered steam, and the Telecom Regulatory Authority of India (TRAI) helped fuel debate in April 2015, requesting public comments on a range of issues related to "Over the Top" services; TRAI expressed concerns about zero-rating as a practice without mentioning Internet.org by name.[67] By early May, TRAI had received over a million public comments, though mostly short ones.[68]

Thus, after several largely uncontested rollouts in other countries, the Internet.org app had run into a wall of opposition in India. When one of the world's biggest technology companies has to change its practices, publically, in response to resistance in one of the world's largest mobile markets, that's a watershed moment. In early May 2015, Mark Zuckerberg, the CEO of Facebook and the most central champion of Internet.org, posted to his blog to report on changes to the Internet.org policies. Instead of exclusively curating a set of pro-development services to accompany a lightweight version of Facebook in the Internet.org app (on Android) and website (on feature phones), Internet.org would be open to any developer as a *platform*, "enabling anyone to build free basic Internet services to help connect the world."[69]

The emphasis on *basic* is important to keep in mind: the guidelines on the platform released the same day as Zuckerberg's blog post suggested that since the services had to work for feature phone users, the free basic web pages seeking inclusion on the platform could not include the relatively more secure HTTPS protocol, use the popular coding language JavaScript, or offer high-resolution photos, video, VOIP, or file transfer.[70] Much like an app store, or Facebook itself, the platform still had several restrictions.[71]

Perhaps this will all be settled by the time this book comes out. If TRAI decides to regulate zero-rating in India, it could rapidly change the climate for zero-rating other markets. However, I rather doubt that the issue will be settled soon. Even the policy research group LIRNEasia, which I have cited elsewhere in the book, cautioned TRAI against making sweeping regulatory recommendations at this point, particularly ones that would ban zero-rating outright.[72] Instead, read the rest of this section as a reflection on the general intersection between zero-rating and structural circumscription, rather than a specific assessment of Internet.org or the India case.

Zero Rating and Net Neutrality

At least prior to the India fracas, zero-rating made great copy for advertising and public relations, and seemed to resonate with resource-constrained consumers. Indeed, once services are in the marketplace, they can be hard to take away, as Blackberry found out when it "irked" customers in South Africa after reducing data subsidies for its BBM messaging service.[73]

In a general sense, some critics suggest that zero-rating undermines the principle of net neutrality,[74] since not all bits exchanged with the ISP or MNO will incur equal treatment (e.g., pricing).[75] Seeing it as a form of vertical price discrimination,[76] some find the idea of zero-rating "absolutely inappropriate"[77] or "a wolf in sheep's clothing,"[78] and not in alignment with the character of the Internet. Others, like Steve Song, have been more measured, noting that net neutrality should be reframed as "a full-stack problem," and pointing out that if the markets for international bandwidth, backhaul, and spectrum are not competitive, it might not be fair to do too much hand-wringing about any one particular element of pricing.[79] At a 2014 meeting of the Internet Governance Forum (IGF), at a session on Zero Rating, Helani Galpaya, CEO of LIRNEasia, was more blunt (note this is a rough real time transcript of the meeting):

Why is [Internet access] unaffordable? It's because there are key parts of the broadband or Internet data chain, the physical data value chain that are uncompetitive. The international backbone for most countries, the prices are very expensive . . .

. . . The lack of competition is a fundamental issue. So, for years regulators have done a poor job, even if the networks are rolling out, of actually making sure enough people are getting connected, i.e., enough people can afford to get connected. In light of this, a new service like the zero-rated content comes in, I think it's a little rich for even regulators and everyone to jump on the bandwagon and say, well it is a holy principle. Yes, the principles are holy but the facts are that our people are not online.[80]

Stepping away from the generalities to practicalities, it is also worth assessing which institutions have been successful with zero-rating, so far, and why. The cases described earlier—Facebook, Google, Twitter, and Wikipedia—are different. The first three create revenue from advertising. Wikipedia creates no revenue but is the symbolic face of what is great about the open, collaborative Internet. Wadhua and Fung from the Wikimedia Foundation talk about the choices leading up to their zero-rating project:

Before launching the program, the Wikimedia foundation (the nonprofit that hosts and operates Wikipedia) took a year gathering feedback and in formally garnering support from the communities that could potentially benefit from it. The opposition to Wikipedia zero came primarily from strong net neutrality advocates that believe no content or services should be made freely available over anything else, even if it's non-commercial or considered by many to be a public good. But from the Wikipedia communities that were actually affected by this program—people with little economic means from countries, such as Kenya and Bangladesh, there was no opposition. In fact, the support was so strong that it grew the community.[81]

Facebook (both on its own and acting under the Internet.org umbrella) and Google have the resources and influence to negotiate dozens of individual deals with MNOs in countries around the world. Wikipedia, although a much smaller organization, has been able to leverage its clout and through leadership, cajoling operators into offering Wikipedia Zero in exchange for the PR, goodwill, and cache which comes with promoting Wikipedia as a free service.

Beyond these examples, many other permutations of fragmented, specified pricing might also take hold. For example, the browser software company Opera offers "Web Pass" service that allows users on participating MNOs to purchase mobile data for specific apps, like "one hour of Facebook" or a "day of Internet"[82]; the U.S. MNO FreedomPop[83] and the startup Pryte (acquired by Facebook)[84] offer similar functionalities. In the United States, the MNO Sprint now offers monthly data plans that include only Facebook and/or Twitter.[85]

Bundled Zero Rating and Editorial Control

Without enforced per-byte pricing equivalence, certain powerful platforms can use these pricing levers to act as curators and editors. Facebook's Internet.org app and website take zero-rating up a level to *bundles* of services, rather than one at a time. Bundled zero-rating has placed the power to select and propagate individual websites with Internet.org (and with the MNOs agreeing to zero-rate it), rather than with users and content providers. (See the participation guidelines for more details on this review process; the process is still managed by Facebook, and "Submission and/or approval by Facebook does not guarantee that your site(s) will be made available through the Internet.org".)[86] Even as a "platform," Internet.org uses the power of zero-rating to introduce bottlenecks of control into the website approval and propagation process; it prenegotiates zero-rating in bulk with

MNOs, and offers to privilege websites playing by its rules over those that will not. To the individual developer, the topic of much of this chapter, the choice is still to run with the juggernaut or risk marginalization trying to go it alone. To individual users, some other group is still curating and dictating which web pages they are more likely to see. These limits on developers and on indivduals are examples of the structural circumscriptions that the more mobile Internet seems to invite. Bandwidth is expensive, and so is attention. Data about users and their habits are useful, and valuable, as well. Any platform that can sit at the nexus of bandwidth subsidies, curational prominence, and extensive data-gathering on user behaviors is likely to be powerful indeed.

ICT4D researcher Mike Best puts it this way: "Does a Facebook on-ramp, even offered for free, describe an available, accessible, and affordable Internet? Facebook is neither a neutral nor open communication platform; it is a business with an architecture designed to support its business plan. This is not a criticism; it is just a reality."[87] Discussing the case of Myanmar, which is just opening up to mobile telephony (and the rest of the world), Ethan Zuckerman goes further: "The Internet in Myanmar is Facebook"; the "future in which Facebook is the Internet of the developing world looks really scary . . . and zero-rating is probably the easiest way for Facebook to get there."[88] To be clear, however, there is not a consensus among the policy communities on Internet.org's approach to zero-rating, just as there is no consensus on the practice more generally. Others have argued that zero-rating is an important and encouraging way to allow resource-constrained users to get online[89]—that some access is better than no access, withheld on principle.

Zero-rating is contentious enough that a few countries, like Chile, the Netherlands, Norway, and Slovenia,[90] have moved to ban it. We will wait to see what TRAI, in India, will do. If other countries do not follow their lead—if zero-rating is here to stay—then those interested in ICT4D and digital inclusion should view it as part of the piecemeal array of solutions to the pricing challenges currently inherent in mobile cellular data. My particular perspective is that it is not zero-rating and subsidized consumption, per se, that is to be avoided, but rather the concentrations of editorial and curatorial power that might result if it is embraced in the wrong way. It is not surprising that major international organizations (both private sector and not-for profit) would seek to zero-rate their content, given the wherewithal and the regulatory latitude to do so. But what about a tiny community message board serving an isolated valley in the Andean foothills? Or a small independent movie studio wanting to release promotions of its new documentary? This is the crux of the power imbalance inherent with the current zero-rating, and why I include this case as one of the ICT4D-specific manifestations of the four circumscriptions discussed in the chapter.

It would be useful for MNOs, companies, regulators, and civil society organizations to explore ways of allowing *any* digital service or content to have the same chances of being zero-rated, on any mobile network. Such a system would put smaller institutions on more equal footing with those with more prestige and bargaining power, and would shift zero-rating from being an incentive used to attract eyeballs or goodwill, toward instead serving as part of a more varied approach toward interactive content delivery. Steve Song has one such proposal along these lines, albeit just a proposal at this stage, to "generically zero-rate 2G services" across the board.[91] The evolution of Internet.org away from a curated collection of apps, toward a broader platform of web pages, is encouraging indeed, with the caveats that several parts of what many would consider "the Internet"—video, high-resolution photos, some security from HTTPS, and so on—are not included in the platform. In the meantime, Jana.com, an Internet startup, has begun to offer developers and advertisers a chance to zero-rate their apps on the networks of over 200 MNOs around the world.[92]

The bottom line, however, is that given the economics and physics of the current mobile cellular wireless infrastructure, someone has to pay for the bandwidth. It can be the MNOs, who can agree to zero-rate a site like Wikipedia in exchange for goodwill and PR; it can be a new zero-rating bundler, like Internet.org; or it can be an advertising platform like Jana.com (arranging to pass the bandwidth costs on to advertisers and app developers). Free to the user does not mean it is free to the system.

A comparison to television is not perfect, but may be apt, given the scarcity of spectrum at play. In TV, a mix of models, including advertising-supported channels broadcast "free" over the air, cable subscriptions, pay-as-you-go on-demand services, and state-supported public television, all seem to have a place. Perhaps the mobile Internet requires a similar diversity for pricing and delivery models for apps and content, and, if so, a greater array of open systems for zero-rating could help bring that about.

Conclusion

When writing this chapter I was reminded of an online (blog) discussion about the shift to mobile Internet between representatives of two great research institutions. On the one hand, Linnet Taylor, a scholar from the Oxford Internet Institute, offered comments drawing on both the interface and generativity themes to suggest that the mobile Internet "wasn't enough": "The mobile Internet is great. But it can't be treated as an end in itself. On even the best smartphone you can't develop software, build anything new, or even produce meaningful amounts of content. You can use, but you can't generate. Without the power to generate content and code, how is the next billion supposed to help develop the web and make it representative?"[93]

Responding directly to this post, Rohan Samarajiva, at LIRNEasia in Sri Lanka, replied with two arguments, along the line of some connectivity is better than no connectivity. One was a caution about extrapolating future states from current technologies. The other was about overgeneralizing use cases beyond the specialized case of producers and academics: "The Internet is a metamedium, the most complex medium ever developed by humans. To expect everyone to use it the same way is wrong. There have to be as many ways of using the Internet as there are people, and perhaps more."[94]

Samarajiva is quite correct about the diversity of uses. Yet the past two chapters have indeed spent considerable time on the special cases of information production. I am not suggesting that everyone will code or be a content creator, or even wants to do so. These questions around production and generativity are significantly broader than "mobile," and better mobile affordances alone are not going to change everything about the global economy or people's places within it. That said, that does not mean that those interested in ICT4D and digital inclusion should not be working to reduce barriers to coding, creating, editing, and distribution erected by the political economy of the more mobile Internet. The few illustrative examples in this chapter should plant seeds of concern in the minds of readers. For more extensive treatment of these issues, I recommend the report from a gathering of researchers at a Bangalore session on inclusion in 2014: "'Inclusion' in digitally-mediated networks," they suggest, "may not [be obtained] with connectivity. Understanding inclusion calls for a grasp over the workings of power, whether connectivity brings control over the terms of participation."[95]

Whenever anyone celebrates how open, flat, inclusive, or decentralized a mobile "solution" or application might be, it is important to evaluate that claim or assumption critically, and skeptically—perhaps especially if that solution seems to be "free." Take the time to assess where power resides in the phenomenon under observation or discussion—not just with the user, or his device, but also in the networks and layers of digital infrastructures supporting them—and then consider whether the enthusiasm is warranted.

IV Implications for Practice and Theory

12   Conclusion

The two previous parts of the book have presented an array of new potentialities and persistent constraints (for development and digital inclusion) associated with the shift to a more mobile Internet. Together with the six elements of mobile Internet experiences described in chapter 4, this array constitutes what I call the After Access Lens. This chapter will revisit the lens as a whole, reflecting on how it might be useful as a guide for further theory and practice in ICT4D, as well as on some elements and issues the lens leaves out.

The After Access Lens, Recapped

Part I—The Boom

After an introduction situating the book as a multidisciplinary effort at the intersection of mobile communication studies and ICT4D, most of part I wrestled with questions of definition and enumeration. Chapters 2 and 3 detailed how the variety of technologies, form factors, and use cases made identification of a clear line between "mobile" and "fixed" Internet very difficult. It is better, I argued, to forget that there is any such thing called "the mobile Internet"; and instead to focus on how the Internet itself is becoming more mobile over time. The world is in the midst of a shift toward a more mobile Internet.

Hence the title *After Access*, since the growth in access has been so rapid and pervasive that it covers the 85 percent of the world's population living under a signal. However, the critical caveat is that 15 percent of the world's population still lives outside the range of any mobile signal, and a far higher percentage suffers from poor-quality signals, expensive tariffs, or prohibitively expensive or difficult handsets. Other technologies like community Wi-Fi and even satellites and drones may help augment the coverage afforded by mobile networks, but, clearly, much work remains to finish the access revolution. I detailed some of that ongoing work—through frugal hardware and software, through expanding coverage and increasing affordability, and through alternative networks—in chapter 2.

Chapter 4 built on the idea of this shift, suggesting it was possible to identify a small set of elements (inexpensive devices, usage-based pricing, wireless connections, personal/portable/intimate devices, universal appeal, and task-supportive experiences), which collectively differentiate mobile Internet experiences from a previous "fixed" Internet archetype. This set contrasted fixed and mobile use cases as antipodes, with innumerable points and paths between them. At that same time, the approach allowed the six elements to play dual roles, both as ways that mobile is different than fixed, and as drivers of the shift to a more mobile Internet. Indeed, the most important contribution of these six elements to digital inclusion and socioeconomic development is the radical step change in access that they support. We have an Internet with two or three billion (perhaps soon to be four billion) users, rather than one or two billion, only because the Internet has become more mobile. It is only through mobile technologies that the Internet has become pervasive, everyday, and inexpensive enough to be truly global and, thus, it is only through mobile technologies that many people have been able to use the Internet for anything at all.

Yet this more mobile Internet is different than the one we might have predicted, before mobile technologies became widespread. These same six mobile elements, individually and collectively, have significant implications for the contours of digital inclusion and socioeconomic development. I concluded the chapter (and thus the section) by presenting an array of implications as the After Access Lens. The lens, with its combination of new mobilities and new constraints, would structure the rest of the book.

Part II—New Mobilities
Part II, "New Mobilities," focused on two new potentialities associated specifically with the shift to a more mobile Internet. Chapter 5, "Places to People, Switching to Serving," began with a careful separation of the social/structural implications of telephony (place-to-place switched connections), mobile telephony (person-to-person switched connections), and Internet

connectivity (read-write access to servers at a distance). With those identified, what is new and unique to mobile Internet experiences is easier to see. Though a mobile handset offers its users *all* these benefits, unlike the other technologies discussed in the chapter, it also offers new potentialities that intermingle place and servers in new ways: place(less) and place(full) informational interactions.

Chapters 6 and 7 explored the implications of these new potentialities: place(less)ness and place(full)ness for ICT4D. Discussing place(less)ness I used examples including gender and agency, connecting diasporas, learning on the move, improved markets, and mobile money. Discussing place(full)ness I described implications for the shape of protests and pressure, and for volunteered geographic information. Each of these domains, and several more I did not cover, open new opportunities for ICT4D practice to help people reconfigure their relationships with each other and with the social and physical spaces in which they live, learn, and work.

Part III—New Constraints

Part III turned from potentialities to constraints, drawing again on the six core elements of mobile communication devices. To do so, chapter 8 introduced two existing concepts from the literature: digital repertoires from media studies and human–computer interaction, and effective use, from community informatics. These two, in combination, allowed us to focus on "mobile-only" and "mobile-centric" digital repertoires (where repertoires are the combination of available technologies and the skills and literacies to use them), and to judge how some repertoires, *ceteris paribus*, may allow more effective use than others. Being mobile-centric is nice, and gives users all the new potentialities described in part II, but for certain times, tasks, and needs, it is still helpful to have access to PCs and unlimited Internet connections.

The remaining chapters in part III introduced three interrelated constraints: a metered mindset, restricted production scenarios, and circumscribed structural roles. Each presents a thicket of new tradeoffs that have come along with the shift toward a more mobile Internet. These three constraints will not affect everybody, all the time, nor is any so bad that we should decry or try to roll back the shift to a more mobile Internet. Yet each reflects and reinforces broader stratifications in the global information economy, and each should be of concern to those who would declare the digital divide "closed" or would seek to use mobile technologies in the service of ICT4D.

In each of these chapters, I identified a mix of priorities to address these persistent stratifications. To mitigate the metered mindset, technical efforts can help users manage their data consumption and support alternative, ideally nonmetered network connections. Data can also be subsidized via "zero-rating." To mitigate restricted production scenarios, efforts should continue to improve production on low-end mobile devices, to train users for increased literacy in mobile-specific informational production, and to convert mobile-only users to mobile-centric users by supporting public access to shared computing resources.

Mitigating the power inequities inherent in the shift to a more mobile Internet is unlikely to be a technical fix, though open source software is a good start. Instead, the power inequities identified with the After Access Lens help refute the propositions some make that the digital divide is closed simply because more people have mobile devices, and indeed suggest that rather than constraints being a function exclusively of the devices, they are bound up with the economics of wireless networks and the business models that have emerged to address them.

To conclude, the After Access Lens is a way to account for a variety of interconnected implications for ICT4D and digital inclusion, coming from the boom in mobile coverage around the world. In all, the shift to a more mobile Internet has brought *new spatial-temporal potentialities* and *new forms of digital stratification*.

Limits to the Lens

The After Access Lens is intended to be flexible, multidisciplinary, and applicable across contexts. As discussed in chapter 4, it is a lens, not a model or a theory. Yet with these intentions come tradeoffs and limits to the lens.

Most importantly, we need to be careful not to attribute more explanatory power to the lens than it actually brings. Conventional literacies, language, and organizational, structural, linguistic, and cultural factors will continue to play a greater role in influencing how individuals and communities use the Internet than will the devices, networks, and services designed to facilitate that use. All the forms of physical, digital, human, and social resources that Warschauer suggests influence "effective use" of ICTs[1] are still at play during this shift to a more mobile Internet. The After Access Lens simply offers a way to bring additional considerations of the increased heterogeneity of device and infrastructure affordances and skills into the ongoing research, policy, and technology discussions around access, the digital divide, and use.

Indeed, with this emphasis on multidisciplinary synthesis and integration comes a demand for frames that can work across multiple geographies, in multiple contexts, and among a user base as diverse as the planet itself. The lens, therefore, did not offer "thick descriptions" of ICT use situated in specific cultural contexts or moments, as one might uncover using the methods of ethnographic research or participant observation. In this way, the lens may instead be useful as a complement to social and anthropological approaches that stress diversity and the co-construction of technologies, users, cultures, and systems. There are no substitutes for these deep dives into context, whether at the macro, national level (e.g., a "Trinidadian" Internet,[2] particularly "Indian" mobile phones),[3] or the micro level, such as exploring the particularities of rural use in remote Papua New Guinea.[4] Similarly, the lens needs to work with, but cannot replace, work and perspectives from other disciplines. For example, the lens might be a takeoff point for explorations of price elasticity for mobile data (microeconomics and policy), or for exploring new topics in mobile-mediated collective action (political science). I have drawn on elements of many disciplines in the aggregation of the lens, but have not subsumed or replaced any of them.

With its focus on "development" and "digital inclusion" the lens has drawn on literatures with a preponderance of studies and pilot projects from sub-Saharan Africa and South Asia, to the relative underweighting of other parts of the Global South. Some of this is probably due to the availability of English-language scholarly publications about these regions, and some to where my own research and networks have been situated. Nevertheless, there are more people using data-enabled mobile devices in China than anywhere else,[5] and thus China is particularly underrepresented in the examples throughout this book. A number of writers have addressed mobile use in China,[6] including Jack Qiu,[7] Cara Wallis,[8] Elisa Oreglia,[9] and Pui-Lam Law,[10] but the links between socioeconomic inclusion and the mobile *Internet* (as distinct from the mobile in general) are at times difficult to tease out. I hope researchers will find the After Access Lens useful when applied to other geographies—it should apply to some degree wherever economic constraint and mobile-only use cases are also prevalent—but only time and further research will tell for sure.

The development and inclusion frame has kept me from any detailed exploration of the *cultural* impacts of a more mobile Internet. From language structure[11] to aesthetics (digital stickers, square photos, and short films, anyone?) to ideas about personal technology use by individuals, institutions,[12] families, and youth,[13] I have no doubt that the "mobile society" literature will continue to wrestle with these potential transformations. That said, the general themes I develop in chapters 6 and 7 of place(full) and place(less) information activities are applicable beyond the purposes of inclusion and development.

The development frame in the book has stressed inclusion and productivity, with considerable focus on livelihoods and expression. This again is a choice I have made, and in so doing have spent less time foregrounding other development frames, such as poverty alleviation, justice, health, well-being and happiness, and environmental sustainability. From phones made with conflict minerals and later turned to e-waste, to the carbon footprints required to run server farms and cell towers,[14] there are issues to consider in a shift to a more mobile Internet that require attention from other researchers.

There are also issues looming that have to do with the arrival of a world where sensors and connected devices vastly outnumber individuals. This is sometimes called the "Internet of Things,"[15] comprised of tens of billions of connected devices. Most of these implications lie beyond the scope of this book. However, one bridge between the topics might be the question of the *density* of devices. Hopefully, the shift to mobile in the Global South will help fuel a more geographically and demographically accurate dispersion of signals for big data, which is increasingly critical to international development decision making.[16] However, if forty billion of the total fifty billion connected devices in the coming Internet of Things are in the developed world, what will the contours of the world that is sensed and connected look like? Similar to the views of lights on Earth at night from space that always have privileged the cities of the prosperous Global North, the places that are sensed (whether mobile or not) may be similarly skewed. The shift to mobile may reduce this differential density, but is unlikely to eliminate it in the near term.

Finally, it is worth stressing one last time that this was not a discussion about phone calls, or person-to-person mobile connections on the go, or especially about Internet connections in general. Each of these other communicative forms may actually prove to be more important to development than the new potentialities of place(full) and place(less)ness that are unique to a more mobile Internet, but to approach all of these elements in a conceptually simultaneous analysis would require a different book, or series of books, or author.[17] The After Access Lens is not a theory of all forms of mediated communication in the developing world; it is simply a way to identify ways in which today's mobile-dominated communicative and informational potentialities differ from what has come before.

After Access: Inclusion, Production, and Effective Use

Gradations, and the Approaching End of the Access Divide

After access, it will be even more apparent that access was not the final destination. Of course, as I have cited throughout the book, many researchers have been arguing this for decades, steering us away from shallow obsessions with a digital divide based on access.[18] We should temper the hype and our enthusiasm—the mobile phone is not going to close the digital divide. Yet that is not so much because the device is flawed, it is because (as I suggested in chapter 3) the utility of the divide concept is so limited.[19] Looking ahead, however, as structural inequities and differences in effective use remain, even as the access divide closes, will ICT4D's theoretical and practical stances be able to account for those differences, and offer coherent, practical advice on how to address them?

Repertoires and Constrained Practice

If anybody with a $10 phone and a little bit of airtime can call up a voice server and manipulate content residing on an Internet server (and then share it with their phone-less friend sitting next to them), the idea of there being a binary divide between those who are connected and those who are not makes little sense. Yet at the same time, it is just as nonsensical to assert that the person with a $10 phone and the person with $1,000 laptop and an always-on broadband connection have the same Internet experience. Thus, the After Access Lens injects additional sources of variability into the concept of Internet use. The lens expands the idea of an Internet user beyond the desktop and the smartphone, to the feature phone user and even to the basic phone user via interfaces like SMS and IVR. These forms do not fit the archetype of decades of accumulated theory about what an Internet experience looks like, but will be common for the next several years before the transition to an all-smartphone installed base is complete.

Part III of the book focused on three general critiques about mobile-Internet use, exploring a mix of conventions (usage-based pricing), technical affordances (around production), and results of broader political-economic factors (of emerging mobile ecosystems). Together, they indicate a variety of interconnected ways in which mobile-only digital repertoires are constrained, relative to repertoires that include a PC connected to the Internet. Evaluating digital repertoires in this way, as opposed to single devices, helps illustrate the "materiality of digital artefacts"[20] at play, and complicates the idea of a binary digital divide that can be closed by mobile devices alone.

Some observers have chosen to describe post-access differences in affordances and use as a form of "digital apartheid."[21] Omitting the word but similar in spirit is an argument that in the United States there are "two Internets" in use.[22] Yet gradations and elements of Internet use described early in this book suggest there are more than two Internets; there are many, some more mobile than others, some offering more expansive potentiality than others.

PC availability will continue to matter in a conversation about mobile Internet use because PCs and mobiles continue to do different things well. Indeed, if access to one device gets people online, then a second, ideally its opposite, may provide a broader range of optimal experiences. People with a wider array of devices spend less time making compromises because they can select the device most suited to their task or desired experience. Conversely, in resource-constrained settings, where personal computers, tablets, standalone video cameras, digital video recording devices, and other technologies are relatively scarce, the data-enabled phone has to do it all, and may be a person's first *owned* digital device.[23] In these conditions, tools of convenience are stretched to serve as tools of necessity. Individuals will need new combinatory literacies and skills[24] to manage not only single devices, but also combinations of devices and the cloud itself[25]—all under conditions of constraint.

Servers, Services, and the Cloud

And yet, the After Access Lens has also revealed ways in which it is increasingly important to run the research gaze beyond the device, toward the services that are running remotely and are used from a mix of fixed and mobile devices. Many new opportunities for promoting ICT4D and digital inclusion are not merely in the continued leveraging of person-to-person communication, but also lie in promoting users' cultivation and control of digital data and identities, as stored on the cloud. From communities mapping their own streets in order to identify service delivery needs,[26] to individual microentrepreneurs establishing credit histories, to teenagers crafting shifting, nuanced, stylized versions of themselves to share online, billions of people are finding it more important to read and write to servers in order to do the stuff of everyday life. As sociologist Jenny Davis says, "We don't have data, we *are* data."[27] Many of these activities have a locative, placeful dimension.[28] And, as those depending on the cloud rise from one or two billion to three or four billion, many—particularly newcomers—will have no choice but to do most of that reading and writing via their mobile devices, on metered connections that discourage extensive data transfer with the cloud.

Production Scenarios

In 2008 and 2009, communication researcher Jack Qiu coined two fantastic phrases to describe the digital repertoires of low-income, mobile-centric ICTs in China. Qiu called their tools "working class ICTs" and called those users "the information have-less."[29] These terms still have relevance and descriptive power today, although perhaps in this case, I might suggest the term "the information *make*-less," since as smartphones have proliferated and Internet access has improved, the remaining shortcomings of mobile-only digital repertoires may have more to do with production, contribution, and control than with consumption or access.

To expand on this centrality of making and production, a new analogy might come from an unlikely place. In a 2010 interview, Steve Jobs offered a powerful vision of a post-PC era by contrasting it to an earlier agrarian era when all cars (on the farm) were trucks: "In the near future, 'PCs are going to be like trucks,' Jobs said. 'They are still going to be around.' However, he said, only 'one of x people will need them.'"[30]

Subsequent years have seemed to support his vision: in both developed and emerging markets, sales of smartphones, tablets, and other post-PC consumer devices are ascendant. Yet in this exact formulation lies the kernel of an implied stratification that must concern ICT4D and digital inclusion theorists and practitioners: *If digital trucks will still do the hard work of production, who will know how to use the trucks, who will drive them? Who will own them, and even design them?*

The After Access Lens suggests that instead of framing this new post-PC era as leaving "1 of x people" requiring a truck, it is rather 1 in x *scenarios* that require digital heavy lifting. These may be fewer, even at home where the PC may sit unused for longer than ever before.[31] The great hope for the next era of computing might be billions of mobiles (at least one for everybody), many tablets (enough for easy sharing, learning and content consumption), and enough bandwidth and PCs (trucks, in Jobs's terminology) to do the heavy lifting for *anyone*.

The Consumer's Mobile Internet Meets the Development Narrative

Despite the heterogeneity of contexts, users, uses, and "apps," serving a billion or more users, a relatively small set of companies play a massive role in structuring the mobile Internet. Some of this is reflective of the Internet in general. As Bruce Sterling put it, the "Stacks" of interconnected, complementary, and layered services offered by each of the largest technology firms are increasingly influential: "[by] 2012 it made less and less sense to talk about 'the Internet,' 'the PC business,' 'telephones,' 'Silicon Valley,' or 'the media,' and much more sense to just study Google, Apple, Facebook, Amazon and Microsoft. These big five American vertically organized silos are re-making the world in their image."[32]

Sterling's language may be a bit dramatic (and it misses some Chinese competitors), but his portrayal may be particularly acute in the case of mobiles: relatively few companies manufacture mobile devices and even fewer create operating systems; networks are provisioned by just a few MNOs per region; and a handful of social networks attract hundreds of millions of users each. Only at the "apps" level is the market considerably fragmented, but even then, chapter 11 described how the stores linked to mobile operating systems hold major editorial and promotional sway over what apps are sold. Concurrently, some of the Stacks provide apps to keep users inside their ecosystems—to serve ads, or mine data, or create better experiences, or cross-promote content, or some combination of all of these rationales. Facebook's "cover all bases" approach in its turn to mobile has been particularly successful.[33] This concentration will continue to affect the course of innovation on the new more-mobile Internet for years to come.

The pressure of these global digital firms, none of which started as telecommunications businesses, will create upheaval in mobile markets. It is not clear how operators will fare as the value of voice and walled-garden value-added services recedes, and the pressure to provide "commodity" access to copious bandwidth increases.[34] Skype (Microsoft) appears on more handsets challenging voice revenues; WhatsApp (Facebook) cuts into SMS messaging. Indeed, Facebook and Google, with enthusiastic forays into efforts like drones and high-altitude balloons, have indicated an interest in reconfiguring the access scenario entirely. In the meantime, in the more prosaic matter of bits and bytes, they are each pushing on operators as both partners for "co-creation" of content and services[35] and potential disruptors, using vehicles like zero-rating to incentivize users.[36]

And yet, beyond these firms, the hardware companies, and the MNOs, progress on the inclusion agenda will require efforts from a broader array of actors, including startups, social enterprises, regulators, and NGOs and international institutions. Some, like the ITU, the World Bank, and the UN's Broadband Commission, come from the multilateral tradition. Others, like Internet.org (led by Facebook), the GSMA (representing MNOs), and the Alliance for Affordable Internet,[37] have close ties to industry interests. For the ICT4D discourse as a whole, it is worth repeating a note of caution from Steve Song: "When it comes to the developing world however there is an amazing dearth of critical discussion about the narrative put forward by communication companies."[38]

It is remarkable and important to consider how the agendas around M4D and ICT4D are enmeshed in public-private partnerships. For example, Internet.org (on Facebook's blog) described an educational initiative called SocialEDU in Rwanda: "At its core, SocialEDU addresses five critical barriers to access by bringing together: Free content, Free data, Affordable smartphones, [l]ocalized, social educational experience, [and a] government that supports innovation."[39] The effort might be promising, and might lead to great educational outcomes for the Rwandan students. However, the After Access Lens might be one way to reveal what is left unaddressed in this formulation, and what stratifications may result from the particular combination of efforts. For example, for whom is the data "free"? Is a smartphone a sufficient tool for promoting learning? And what does it mean for the government to support innovation in technology through national-level partnerships with global firms? The point here is not to single out this initiative for critique, but rather to show how the lens, with attention to costs, affordances, and political economies, can quickly raise important questions.

The shift to a more mobile Internet influences one additional tension in the ICT4D community, between models and perspectives that foreground appropriation and user choice of general technologies, versus those that favor intervention and design with or on behalf of a target community with a specific goal of improving that community on some set of development-related outcomes. Sometimes, this boils down to choices of acronyms, between an open-ended, sometimes passive, often vague ICTD and a more intentional, interventionist ICT4D. Despite the occasional "4D" rhetoric from groups like the GSMA and the World Bank, one thing that is striking about the turn to the mobile Internet in the developing world is how little "4D" was actually involved. Facebook, QQ, Sina Weibo, Twitter, WhatsApp, MXit, YouTube, and Google (to name a few) will be as important to the lives and livelihoods of many resource-constrained people as any specialty application flying the ICT4D (or M4D) flag. Is ICT4D theory ready for this? If advertising companies like Facebook and Google are taking the lead in closing remaining access gaps, and are imagining new ways to get the most out of the existing mobile Internet infrastructure, ICT4D needs to understand both the developmental and nondevelopmental impacts of the centrality of these actors.

A theoretical goal is to consider how *these* institutions are thinking about solving users' problems. What kinds of interventions are being enabled? How do they link to the Internet in general, and the idea of an empowered user? To return to the SocialEDU example, what problems does industry imagine it is solving?[40] Further unpacking of these roles, expanding and engaging with the arguments made by Song, Gurumurthy, Zuckerman, Hersman, and others,[41] will push the boundaries of ICT4D, and will illustrate how there are often few clear distinctions between an ICT4D app and a general app that happens to be used for ICT4D. The centrality of nondevelopmental actors does not replace or negate traditional ICT4D interventions, whether in health, education, livelihoods, governance, sanitation, environment, or whatever. Pilot programs still need to be fielded, market failures can be addressed, and people can work tirelessly alongside and with communities to help everyone get more out of the technologies that they have. However, would-be interventionists cannot and should not approach the tasks in a vacuum that does not account for these other actors and for the realities of a relatively narrow set of activities and services that have come to define the mobile Internet for many users.

This book has maintained a relatively narrow focus on improving the tools and skills that could enable (or at least, not further impede) participation in the global informational economy and society. Yet actions involving the "Stacks," including social.EDU and Internet.org examples, may be tantalizing for further, more comprehensive, philosophical, and fundamental critiques of the relationships between technologies and global economies. This is generally well beyond the charge of this book. However, this book, through the After Access Lens, can signal how the shift to mobile may alter some of the terms according to which information haves and have-lesses[42] (to echo Jack Qiu's term) interact in the global information economy.

In a related matter, the book has been reticent in suggesting specific actions national governments could take to address the stratifications identified by the After Access Lens. Mostly, this is because the circumstances vary from country to country. I mentioned the different responses to zero-rating in the last chapter. More generally, government action could manifest in the United States in the form of public support for digital public media;[43] in South Africa, it could be support for public access.[44] Brazil is a fascinating model to consider, having passed the Marco Civil da Internet (The Internet Bill of Rights).[45]

As a frequent participant in the ICT4D discussions, I sometimes worry if the community as a whole is sufficiently engaged with these deeper structural issues. Some theorists and practitioners in ICT4D certainly are engaged in this way. There is, at times, a palpable tension in the ICT4D community between critical perspectives offering critiques of neoliberalism, structural inequality, and global capital, and perspectives stressing (seemingly) apolitical technocratic improvements to pressing problems. I would argue that the shift toward widespread mobile Internet access is going to force more dialogue between these perspectives within ICT4D. Building new cell towers and inventing new functionalities may indeed reduce poverty and facilitate some inclusion, but if structural inequalities persist, or if productivity growth stalls, or both, ICT4D will need updated narratives.

In particular, a theoretically rigorous response to the off-the-cuff comment from a leader in industry that "PCs will be like trucks" might be helpful. However, deeper and more explicit linkages that move beyond access to understand ICTs *as the means of production*—of both economic value and cultural meaning—will be critical for the next phase of ICT4D. Perhaps ICT for development should become ICT for production. Mobile-only repertoires support coordination, consumption, and some circumscribed means of informational production. Yet this book suggests that ICT4D needs to be careful about over-attributing productive capacities to the mobile-only repertoire. In an age when computers are "culture machines," I echo Lunenfeld's description of the "secret war between uploading and downloading."[46] Smartphones are also culture machines, but they are not the *same* machines as personal computers. The smartphone has been optimized to support the on-the-go and on-demand needs of the post-industrial knowledge worker and hyperconnected social media prosumer. ICT4D needs to build bridges to the technology and society literatures to ascertain if these devices and the services to which they connect are also optimized for the farmers, fishermen, and artisans (and their families) that figure so prominently in the development discourse. Mobile-only repertoires can support inclusion and production, but ICT4D needs to ask, on whose terms?[47] Are those with mobile-only digital repertoires in control of their digital data? Do they have full read-write access to the information and services they need?[48]

The End of M4D
As I mentioned in chapter 1, the "mobiles for development" term "M4D" is popular in practitioner communities, and appears in a small but growing set of research articles,[49] including in a few authored by me. However, the ideas of digital repertoires, heterogeneous device environments, and the After Access Lens challenge the boundaries and utility of this term. M4D may have made sense in the earlier days of the mobile telecommunications boom in the Global South, when there was little functional overlap between the networks and functionalities of the mobile phone and the Internet. "M4D" could stand, for a time, as the shorthand contrast to the strains and frustrations surrounding the Internet thesis (aka "ICT4D"). Mobile had the momentum of adoption and simplicity behind it.

Through this book, and particularly chapters 3 and 4, it should be apparent how the lines between devices, networks, outcomes, and uses have become so blurred as to make the distinction between ICT4D and M4D difficult to discern. Instead, interested parties should assess, discuss, and utilize different devices and networks used to access and manipulate the same data residing on servers and the cloud, albeit with significant and consequential differences in functionality, political economy, and affordances. Nearly any attempt to understand or improve the use of ICTs for socioeconomic development or inclusion needs to shift to a frame that foregrounds material multidevice user repertoires, and, equally, immaterial cloud services and infrastructures to which those devices connect.[50] Neither of these shifts is facilitated by the persistence of a subcommunity flying the flag of a single device+network modality; to be blunt, it is time to retire the term "M4D."

Conclusion

I should conclude with some recalibration; readers deserve some reassurance that the general story remains positive. Thanks to the worldwide cellular infrastructure, a boom in Internet connectivity is well underway. In most cases, *some* Internet access is better than *no* Internet access, a *somewhat* affordable Internet is better than an unaffordable Internet, and an Internet that is easy and fun to use is better than one that is difficult. I have dedicated relatively little space in this volume to a discussion of the positive (or negative) impacts of the Internet in general, as used via the mobile, for outcomes in health, education, participation, prosperity, and so on. There is a tremendous amount to be excited about, thanks to the spread of a more mobile, accessible Internet.

Instead, the After Access Lens concentrates on how a more mobile Internet presents new potentialities, in addition to what has come before via phones, mobiles, and fixed Internet connections. If I may be permitted a bit of flourish in the last pages of the book, these potentialities are new superpowers people can add to the already-nearly-magic capabilities to share thoughts across distances at the speed of light, and to access the world's information without leaving one's home.[51] Place(full) and place(less) interactions with servers and systems may not yet be as important to ICT4D outcomes as a phone call, or a basic Internet connection, but they are fascinating, and promise changes to the social-spatial organization of communities and states in the Global South that we are just beginning to see.

That said, part III of the book identified work that still needs to be done to make the mobile Internet more useful and usable to more people around the world. Using the concepts of digital repertoires and heterogeneous device environments, I illustrated how the same properties of mobile Internet experiences that have been driving its rapid growth are also reinforcing and introducing persistent differences (stratifications) in Internet experiences. The Internet experience for those with mobile-only digital repertoires is not the same as that experienced by those who can reliably access and effectively use PCs, especially over unmetered data connections.

I would like to stress again that other researchers[52] are engaging with the implications of these stratifications for policy and development practice. Marion Walton[53] has published extensively on mobile-centric Internet use from a media studies perspective, as has Katy Pearce with her colleagues in communication research.[54] Some prominent activists and development practitioners, including Ethan Zuckerman,[55] Steve Song,[56] and Roger Harris[57] have raised concerns about the limitations of the centralized, costly mobile channel as the exclusive means by which many will access the Internet. The Pew Research Center[58] has been the source of valuable metrics on mobile-centric Internet use both in the United States and internationally. Also in the policy arena, papers by Horner[59] and by Napoli and Obar[60] caution against declaring the digital divide "closed" while Samarajiva and his colleagues at LIRNEasia in Sri Lanka are more enthusiastic about "more than voice services" and the promise of the "budget telco model"[61] as applied to the mobile Internet. All of their work has been influential in the framing of this book.

In the journal *Telecommunications Policy*, Forestier, Grace, and Kenny asked readers to focus less on a debate about whether telecommunications are beneficial to humanity, and more on the complex relationships between progress and fairness: "We know that telecommunications is pro-poor to the extent that it promotes growth, but it might be 'super pro-poor' if it promotes growth while increasing equity. Conversely, it might be 'sub pro-poor' if it increases growth at the cost of equity."[62]

I offered these inquiries into digital repertoires, mobile-only Internet use, heterogeneous device environments, and effective use in the spirit of identifying actions that can coax the mobile Internet toward being super pro-poor. *After Access* has synthesized disparate threads across a variety of literatures, and challenges an emerging truism that mobile telephony is the cure-all for the digital divide and global inequity. A broadly applicable policy and design case can be made for improving resource-constrained digital repertoires, making them more useful and inclusive. Research can help inform decisions about design and deployment of mobile ICTs, about pricing them and their networks, as well as methods for training people to use them. In combination, these efforts would improve the digital repertoires of those who rely on mobile devices as their primary or exclusive ways of accessing and using the Internet.

I have made these arguments without offering a stance as to whether access to the Internet is a human right, a basic need, or a public good. The United Nations Human Rights Council took some (admittedly nonbinding and unenforceable) steps in that direction in 2012, calling "upon all States to promote and facilitate access to the internet and international cooperation aimed at the development of media and information and communications facilities in all countries."[63] That said, I like the formulation of "Internet Justice" that community informatics scholar Mike Gurstein has proposed—a vision that "the Internet is and continues to be a resource available, usable and of equitable benefit to all."[64] The shift to a more mobile Internet brings new challenges for Internet justice.

The Internet that will envelop and engage the majority of the planet's inhabitants is not the one that we have theorized about for twenty-five years. Access and use of the Internet via mobile technologies offers personal, convenient, and pervasive experiences, including ones taking advantage of new potentialities of place(full) and place(less)ness. However, the systems, as currently deployed, are expensive for many resource-constrained people to use on an ongoing basis; their affordances remain better for coordination and consumption than for intensive informational production; and the underlying technologies are resistant to open tinkering and adaptation.

Granted, having even more accessible informational production technologies will not solve all the world's development challenges.

Nevertheless, a broader, more mobile Internet, used effectively by more people, creates great opportunities for individuals to self-organize, to coordinate, to learn, and to reshape their communities in ways that may lead to greater happiness, wellness, prosperity, and agency. A focus on repertoires and effective use, instead of on successive waves of discrete artifacts, will help the ICT4D field transition to an increasingly post-access world, and focus on the challenges of inclusion and production that still confront it.

Notes

Note to readers: The references below have been gathered in note form, arranged by chapter. An integrated bibliography—a list of all references in the book, in alphabetical order—will be made available around the time of publication at www.jonathandonner.com/afteraccessreferences.

Chapter 1

1. Richard Ling and Jonathan Donner, *Mobile Communication* (Cambridge, UK: Polity, 2009); Valerie Feldmann, "Mobile Overtakes Fixed: Implications for Policy and Regulation" (Geneva: International Telecommunication Union, 2003), http://www.itu.int/osg/spu/ni/mobileovertakes/Resources/Mobileovertakes_Paper.pdf.

2. Quoted in Jack Ewing, "Upwardly Mobile in Africa," *Bloomberg Businessweek*, September 23, 2007, http://www.businessweek.com/stories/2007-09-23/upwardly-mobile-in-africa.

3. Michael Minges, "Overview," in *Information and Communications for Development 2012: Maximizing Mobile*, ed. World Bank (Washington, DC: World Bank, 2012), 11–30, quotation from p. 12.

4. Paul Craven and Barry Wellman, "The Network City," *Sociological Inquiry* 43, no. 3–4 (July 1973): 57–88, doi:10.1111/j.1475-682X.1973.tb00003.x.

5. Stephane Boyera, "Can the Mobile Web Bridge the Digital Divide?," *Interactions* 14, no. 3 (May 01, 2007): 12, doi:10.1145/1242421.1242433; Andrew Grantham and George Tsekouras, "Information Society: Wireless ICTs' Transformative Potential," *Futures* 36, no. 3 (April 2004): 359–377, doi:10.1016/S0016-3287(03)00066-1.

6. For details on the approach taken toward research and development at MSR, see http://research.microsoft.com/en-us/about/.

7. Jonathan Donner, Shikoh Gitau, and Gary Marsden, "Exploring Mobile-Only Internet Use: Results of a Training Study in Urban South Africa," *International Journal of Communication* 5 (2011): 574–597; Shikoh Gitau, Gary Marsden, and Jonathan

Donner, "After Access: Challenges Facing Mobile-Only Internet Users in the Developing World," in *Proceedings of the 28th International Conference on Human Factors in Computing Systems—CHI '10* (New York: ACM, 2010), 2603–2606, doi:10.1145/1753326.1753720.

8. Marsden, Gitau, and I liked the term "After Access," but were not its only fans. It sometimes turns up as a phrase to define post-access digital divides or differences. Yong-Chan Kim, Joo-Young Jung, Elisa L. Cohen, and Sandra J. Ball-Rokeach, "Internet Connectedness Before and After September 11, 2001," *New Media & Society* 6, no. 5 (October 01, 2004): 611–631, doi:10.1177/146144804047083; John P. Robinson, Paul DiMaggio, and Eszter Hargittai, "New Social Survey Perspectives on the Digital Divide," *IT&Society* 1, no. 5 (2003): 1–22. Ari Katz, "'After Access, What?' Reflections Following Salon Discussion in the Philippines," *Beyond Access* (blog), accessed November 11, 2014, http://beyondaccess.net/2012/05/17/after-access-what -reflections-following-salon-discussion-in-the-philippines.

9. Jonathan Donner and Marion Walton, "Your Phone Has Internet—Why Are You at a Library PC? Re-Imagining Public Access for the Mobile Internet Era," in *Proceedings of INTERACT 2013: 14th IFIP TC 13 International Conference* (Berlin: Springer, 2013), 347–364, doi:10.1007/978-3-642-40483-2_25.

10. Jonathan Donner and Andrew Maunder, "Beyond the Phone Number: Challenges of Representing Informal Microenterprise on the Internet," in *Living Inside Mobile Social Information*, ed. James E. Katz (Dayton, Ohio: Greyden Press, 2014), 159–192.

11. Preeti Mudliar, Jonathan Donner, and William Thies, "Emergent Practices around CGNet Swara: A Voice Forum for Citizen Journalism in Rural India," *Information Technologies & International Development* 9, no. 2 (2013): 65–79; Preeti Mudliar and Jonathan Donner, "Experiencing Interactive Voice Response (IVR) as a Participatory Medium: The Case of CGNet Swara in India," *Mobile Media & Communication*, published online before print (2015), doi:10.1177/2050157915571591.

12. Jonathan Donner, "Research Approaches to Mobile Use in the Developing World: A Review of the Literature," *The Information Society* 24, no. 3 (2008): 140–159, doi:10.1080/01972240802019970.

13. Ricardo Gomez, Luis F. Baron, and Brittany Fiore-Silfvast, "The Changing Field of ICTD," in *Proceedings of the Fifth International Conference on Information and Communication Technologies and Development—ICTD '12* (New York: ACM Press, 2012), 65–74, doi:10.1145/2160673.2160682; Tim Unwin, *ICT4D: Information and Communication Technology for Development* (Cambridge, UK: Cambridge University Press, 2009).

14. Everett M. Rogers, "Communication and Development: The Passing of the Dominant Paradigm," *Communication Research* 3, no. 2 (April 1, 1976): 213–240, doi:10.1177/009365027600300207.

15. Anita Gurumurthy, "From Social Enterprises to Mobiles—Seeking a Peg to Hang a Premeditated ICTD Theory in an Ecology of Unequal Actors," *Information Technologies & International Development* 6, special issue (2010): 57–63; Don Slater, *New Media, Development & Globalization* (Cambridge, UK: Polity Press, 2013).

16. FHI 360, "Mobiles! What Have We Learned? Where Are We Going?" (Washington, DC, September 26, 2013), http://mobiledevconference.com/.

17. John Sören Pettersson, ed., *Proceedings of the 1st International Conference on M4D Mobile Communication Technology for Development*. December 11–12, 2008, Karlstad University, Sweden (Karlstad: Karlstad University Studies, 2008).

18. World Bank, *Information and Communications for Development 2012: Maximizing Mobile* (Washington, DC: The World Bank, 2012), doi:10.1596/978-0-8213-8991-1.

19. GSMA Mobile for Development Intelligence, *Scaling Mobile for Development: Harness the Opportunity* (London, August 2013), https://gsmaintelligence.com/research/?file=130828-scaling-mobile.pdf; Vodafone, "Africa: The Impact of Mobile Phones," *Policy Paper Series #2* (London, March 2005), http://www.vodafone.com/content/dam/vodafone/about/public_policy/policy_papers/public_policy_series_2.pdf.

20. GSMA Mobile for Development Intelligence, *Scaling Mobile for Development*.

21. Jonathan Donner, "Framing M4D: The Utility of Continuity and the Dual Heritage of 'Mobiles and Development,'" *The Electronic Journal of Information Systems in Developing Countries* 44, no. 3 (2010): 1–16; Jonathan Donner, Katrin Verclas, and Kentaro Toyama, "Reflections on MobileActive08 and the M4D Landscape," in *Proceedings of the First International Conference on M4D*, ed. John Sören Pettersson (Karlstad, Sweden: Karlstad University Studies, 2008), 73–83.

22. Arul Chib, "The Promise and Peril of mHealth in Developing Countries," *Mobile Media & Communication* 1, no. 1 (January 2013): 69–75, doi:10.1177/2050157912459502.

23. Jenny C. Aker, "Dial 'A' for Agriculture: A Review of Information and Communication Technologies for Agricultural Extension in Developing Countries," *Agricultural Economics* 42, no. 6 (November 2011): 631–647, doi:10.1111/j.1574-0862.2011.00545.x.

24. John Traxler and Steve Vosloo, "Introduction: The Prospects for Mobile Learning," *Prospects* 44, no. 1 (2014): 13–28, doi:10.1007/s11125-014-9296-z.

25. Darrell M. West and Elizabeth Valentini, "How Mobile Devices Are Transforming Disaster Relief and Public Safety" (Washington, DC: Brookings Institution, July 2013), http://www.brookings.edu/~/media/research/files/papers/2013/07/16-mobile-technology-disaster-relief/west_valentini_mobile-technology-disaster-relief_v20.pdf.

26. Siddhartha Raja, Katherine Maher, Michael Minges, and Priya Surya, "Making Government Mobile," in *Information and Communications for Development 2012: Maximizing Mobile* (Washington, DC: World Bank, 2012), 87–102.

27. Sokari Ekine, ed., *SMS Uprising: Mobile Activism in Africa* (Cape Town, South Africa: Pambazuka Press, 2009).

28. Misha T. Hutchings, Anurupa Dev, Meena Palaniappan, Veena Srinivasan, Nithya Ramanathan, John Taylor, Nancy Ross, and Paula Luu, "mWASH: Mobile Phone Applications for the Water, Sanitation, and Hygiene Sector" (Los Angeles: Nextleaf Analytics, April 2012), http://pacinst.org/wp-content/uploads/sites/21/2014/04/mwash.pdf.

29. Richard A. Duncombe, "Understanding the Impact of Mobile Phones on Livelihoods in Developing Countries," *Development Policy Review* 32, no. 5 (2014): 567–588, doi:10.1111/dpr.12073.

30. Ithiel de sola Pool, ed., *The Social Impact of the Telephone* (Cambridge, MA: MIT Press, 1977); Claude S. Fischer, *America Calling: A Social History of the Telephone to 1940* (Berkeley: University of California Press, 1992).

31. Steve Jones, Veronika Karnowski, Richard Ling, and Thilo Von Pape, "Welcome to Mobile Media & Communication," *Mobile Media & Communication* 1, no. 1 (2013): 3–7, doi:10.1177/2050157912471456.

32. For example, James E. Katz and Mark Aakhus, "Conclusion: Making Meaning of Mobiles: A Theory of Apparatgeist," in *Perpetual Contact: Mobile Communication, Private Talk, and Public Performance*, ed. James E. Katz and Mark Aakhus (Cambridge, UK: Cambridge University Press, 2002), 301–318; Manuel Castells, Mireia Fernández-Ardèvol, Jack Linchuan Qiu, and Araba Sey, *Mobile Communication and Society: A Global Perspective* (Cambridge, MA: MIT Press, 2007); Richard Ling, *New Tech, New Ties* (Cambridge, MA: MIT Press, 2008); Gerard Goggin, *Cell Phone Culture: Mobile Technology in Everyday Life* (New York: Routledge, 2006). I draw extensively on the 2014 *Routledge Companion to Mobile Media* for summaries of many of the concepts throughout the book: Gerard Goggin and Larissa Hjorth, eds., *Routledge Companion to Mobile Media* (New York: Routledge, 2014).

33. Donner, "Research Approaches to Mobile Use in the Developing World."

34. Richard Ling and Heather A. Horst, "Mobile Communication in the Global South," *New Media & Society* 13, no. 3 (March 31, 2011): 363–374, doi:10.1177/1461444810393899.

35. Heather A. Horst and Daniel Miller, *The Cell Phone: An Anthropology of Communication* (Oxford: Berg, 2006).

36. Assa Doron and Robin Jeffrey, *The Great Indian Phone Book: How the Cheap Cell Phone Changes Business, Politics, and Daily Life* (Cambridge, MA: Harvard University Press, 2013).

37. Michael Gurstein, *Community Informatics: Enabling Communities with Information and Communications Technologies* (Hershey, PA: Idea Group, 2000).

38. Regina Roth, "Marx on Technical Change in the Critical Edition," *The European Journal of the History of Economic Thought* 17, no. 5 (December 2010): 1223–1251, doi: 10.1080/09672567.2010.522239; Christopher May, "The Information Society as Mega-Machine: The Continuing Relevance of Lewis Mumford," *Information, Communication & Society* 3, no. 2 (2000): 241–265.

39. James R. Beniger, *The Control Revolution: Technological and Economic Origins of the Information Society* (Cambridge, MA: Harvard University Press, 1986).

40. Manuel Castells, *The Rise of the Network Society* (Malden, MA: Blackwell Publishing, 1996).

41. Barry Wellman, *Networks in the Global Village: Life in Contemporary Communities* (Boulder, CO: Westview Press, 1999).

42. Bruno Latour, *Science in Action* (Cambridge, MA: Harvard University Press, 1987).

43. Vandana Desai and Robert B. Potter, eds., *The Companion to Development Studies* (New York: Routledge, 2013).

44. Jonathan Donner, "Shrinking Fourth World? Mobiles, Development and Inclusion," in *Handbook of Mobile Communication Studies*, ed. James E. Katz (Cambridge, MA: MIT Press, 2008), 29–42.

45. Marco Zeschky, Bastian Widenmayer, and Oliver Gassmann, "Frugal Innovation in Emerging Markets," *Research-Technology Management* 54, no. 4 (July 1, 2011): 38–45, doi:10.5437/08956308X5404007.

46. Christopher Foster and Richard Heeks, "Analyzing Policy for Inclusive Innovation: The Mobile Sector and Base-of-the-Pyramid Markets in Kenya," *Innovation and Development* 3, no. 1 (2013): 103–119, doi:10.1080/2157930X.2013.764628.

47. Renee Kuriyan, Isha Ray, and Kentaro Toyama, "Information and Communication Technologies for Development : The Bottom of the Pyramid Model in Practice," *The Information Society* 24, no. 789921171 (March 04, 2008): 93–104, doi:10 .1080/01972240701883948; Aneel Karnani, "The Mirage of Marketing to the Bottom of the Pyramid: How the Private Sector Can Help Alleviate Poverty," *California Management Review* 49, no. 4 (2007): 90–112.

48. C. K. Prahalad and Allen L. Hammond, "Serving the World's Poor, Profitably," *Harvard Business Review* 80, no. 9 (2002): 48–59.

49. Ashish Karamchandani, Michael Kubzansky, and Paul Frandano, "Emerging Markets, Emerging Models: Market Based Solutions to the Challenges of Global Poverty," *Monitor Group Reports* (Cambridge, MA, March 2009), http://www .mim.monitor.com/downloads/emergingmarkets_full.pdf; Patrick Whitney and

Anjali Kelkar, "Designing for the Base of the Pyramid," *Design Management Review* 15, no. 4 (June 2010): 41–47, doi:10.1111/j.1948-7169.2004.tb00181.x.

50. William H. Dutton, ed., *The Oxford Handbook of Internet Studies* (Oxford, UK: Oxford, 2013), doi:10.109310.1093/oxfordhb/9780199589074.001.0001.

51. Tai-Quan Peng, Lun Zhang, Zhi-Jin Zhong, and Jonathan JH Zhu, "Mapping the Landscape of Internet Studies: Text Mining of Social Science Journal Articles 2000–2009," *New Media & Society* 15, no. 5 (November 2012): 644–664, doi:10.1177/1461444812462846.

52. Genevieve Bell and Paul Dourish, "Yesterday's Tomorrows: Notes on Ubiquitous Computing's Dominant Vision," *Personal and Ubiquitous Computing* 11, no. 2 (November 2006): 133–143, doi:10.1007/s00779-006-0071-x.

53. Jason Farman, "Introduction to the Social Transformations from the Mobile Internet Special Issue," *Future Internet* 4, no. 2 (May 2012): 545–550, doi:10.3390/fi4020545.

54. Harmeet Sawhney, "Innovations at the Edge: The Impact of Mobile Technologies on the Character of the Internet," in *Mobile Technologies: From Telecommunications to Media*, ed. Gerard Goggin and Larissa Hjorth (New York: Routledge, 2009), 105–117.

55. Jack Linchuan Qiu, *Working-Class Network Society: Communication Technology and the Information Have-Less in Urban China* (Cambridge, MA: MIT Press, 2009).

56. Ayesha Zainudeen, "Are the Poor Stuck in Voice? Conditions for Adoption of More-Than-Voice Mobile Services," *Information Technologies & International Development* 7, no. 3 (2011): 45–59.

57. Wallace Chigona, Darry Beukes, Junaid Vally, and Maureen Tanner, "Can Mobile Internet Help Alleviate Social Exclusion in Developing Countries?," *The Electronic Journal of Information Systems in Developing Countries* 36 (2009): 1–16.

58. Peter Benjamin, "mHealth Hope or Hype: Experiences from Cell-Life," in *mHealth in Practice: Mobile Technology for Health Promotion in the Developing World*, ed. Jonathan Donner and Patricia Mechael (London: Bloomsbury Academic, 2012), 62–71.

59. Gary Marsden, Lucia Terrenghi, and Matt Jones, "Globicomp—Doing Ubicomp Differently: Introduction to the Special Issue," *Personal and Ubiquitous Computing* 15, no. 6 (December 14, 2010): 551–552, doi:10.1007/s00779-010-0336-2; Gary Marsden, "What Is the Mobile Internet?," *Interactions* 14, no. 6 (November 2007): 24–25, doi:10.1145/1300655.1300672.

60. Marion Walton, "Mobile Literacies and South African Teens: Leisure Reading, Writing, and MXit Chatting for Teens in Langa and Guguletu," *Report for the Shuttleworth Foundation* (Cape Town, December 2009), https://m4lit.files.wordpress

.com/2010/03/m4lit_mobile_literacies_mwalton_20101.pdf; Marion Walton, "Pavement Internet: Mobile Media Economies and Ecologies in South Africa," in *Routledge Companion to Mobile Media*, ed. Gerard Goggin and Larissa Hjorth (London: Routledge, 2014), 450–461; Marion Walton and Pierrinne Leukes, "Prepaid Social Media and Mobile Discourse in South Africa," *Journal of African Media Studies* 5, no. 2 (2013): 149–167, doi:10.1386/jams.5.2.149_1; Marlon Walton, Gary Marsden, Silke Hassreiter, and Sena Allen, "Degrees of Sharing: Proximate Media Sharing and Messaging by Young People in Khayelitsha," in *Proceedings of the 14th International Conference on Human–Computer Interaction with Mobile Devices and Services—MobileHCI '12* (New York: ACM Press, 2012), 403–412, doi:10.1145/2371574.2371636.

61. Katy E. Pearce and Ronald E. Rice, "Digital Divides From Access to Activities: Comparing Mobile and Personal Computer Internet Users," *Journal of Communication* 63, no. 4 (2013): 721–744, doi:10.1111/jcom.12045; Katy E. Pearce, "Convergence through Mobile Peer-to-Peer File Sharing in the Republic of Armenia," *International Journal of Communication* 5, (2011): 511–528; Katy E. Pearce, Janine Slaker, and Nida Ahmad, "Is Your Web Everyone's Web? Theorizing the Web through the Lens of the Device Divide," *Theorizing the Web 2012*, April 12, 2012, http://fr.slideshare.net/katyp1/katy-pearce-ttw12; Katy E. Pearce, "Phoning It In: Theory in Mobile Media and Communication in Developing Countries," *Mobile Media & Communication* 1, no. 1 (January 2013): 76–82, doi:10.1177/2050157912459182.

62. Nimmi Rangaswamy and Edward Cutrell, "Anthropology, Development and ICTs," in *Proceedings of the Fifth International Conference on Information and Communication Technologies and Development—ICTD '12* (New York: ACM Press, 2012), 85–93, doi:10.1145/2160673.2160685; Nimmi Rangaswamy and S. Yamsani, "'Mental Kartha Hai' or 'Its Blowing My Mind': Evolution of the Mobile Internet in an Indian Slum," in *Proceedings of the 2001 Ethnographic Praxis in Industry Conference* (American Anthropological Association, 2011), 285–298, https://www.epicpeople.org/wp-content/uploads/2014/09/Rangaswamy_menta.pdf; Preeti Mudliar and Nimmi Rangaswamy, "Offline Strangers, Online Friends," in *Proceedings of the 33rd Annual ACM Conference on Human Factors in Computing Systems—CHI '15*, (New York: ACM, 2015), 3799–3808, doi:10.1145/2702123.2702533.

63. Slater, *New Media, Development & Globalization*.

64. Dallas W. Smythe and Tran Dinh, "On Critical and Administrative Research: A New Critical Analysis," *Journal of Communication* 33, no. 3 (September 1983): 117–127, doi:10.1111/j.1460-2466.1983.tb02413.x; William H. Melody and Robin E. Mansell, "The Debate over Critical vs. Administrative Research: Circularity or Challenge," *Journal of Communication* 33, no. 3 (1983): 103–116, doi:10.1111/j.1460-2466.1983.tb02412.x.

65. Jo Tacchi, "Being Meaningfully Mobile: Mobile Phones and Development," in *Technological Determinism and Social Change: Communication in a Tech-Mad World*, ed. Jan Servaes (Lanham, MD: Lexington Books, 2014), 105–124.

66. Huatong Sun and William F. Hart-Davidson, "Binding the Material and the Discursive with a Relational Approach of Affordances," in *Proceedings of the 32nd Annual ACM Conference on Human Factors in Computing Systems—CHI '14* (New York: ACM Press, 2014), 3533–3542, doi:10.1145/2556288.2557185.

67. Emmanuel Forestier, Jeremy Grace, and Charles Kenny, "Can Information and Communication Technologies Be Pro-Poor?," *Telecommunications Policy* 26, no. 11 (December 2002): 623–646, doi:10.1016/S0308-5961(02)00061-7.

68. Charles Kenny, *Getting Better: Why Global Development Is Succeeding—And How We Can Improve the World Even More* (New York: Basic Books, 2012).

69. Castells, *The Rise of the Network Society*. Or, Eubanks's description of the challenges of using technologies for empowerment in upstate New York. Virginia Eubanks, *Digital Dead End: Fighting for Social Justice in the Information Age* (Cambridge, MA: MIT Press, 2011).

70. Michael Gurstein, "Internet Justice: A Meme Whose Time Has Come," *Gurstein's Community Informatics Blog*, November 27, 2013, http://gurstein.wordpress.com/2013/11/27/internet-justice-a-meme-whose-time-has-come/.

Chapter 2

1. Valerie Feldmann, "Mobile Overtakes Fixed: Implications for Policy and Regulation" (Geneva: International Telecommunication Union, 2003), http://www.itu.int/osg/spu/ni/mobileovertakes/Resources/Mobileovertakes_Paper.pdf. This prescient report included reflections on the importance of prepay pricing, and mobile data's potential to spread beyond early successes in Japan and South Korea.

2. International Telecommunication Union, "Key 2005–2014 ICT Data," *ITU-D Statistics Online*, 2014, http://www.itu.int/en/ITU-D/Statistics/Documents/statistics/2014/ITU_Key_2005-2014_ICT_data.xls.

3. Ibid.

4. Roshanthi Lucas Gunaratne and Rohan Samarajiva, "Estimating Internet Users: An Evidence-Based Alternative in the Absence of Survey Data," *Info* 15, no. 5 (2013): 20–33, doi:10.1108/info-05-2013-0023.

5. Ericsson, "Ericsson Mobility Report: On the Pulse of the Networked Society" (Stockholm, November 2014), http://www.ericsson.com/res/docs/2014/ericsson-mobility-report-november-2014.pdf.

6. Mary Meeker, "2012 Internet Trends @Stanford–Bases," December 3, 2012, http://www.kpcb.com/file/kpcb-2012-internet-trends-update.

7. Ewan Sutherland, "Counting Customers, Subscribers and Mobile Phone Numbers," *Info* 11, no. 2 (2009): 6–23, doi:10.1108/14636690910941858; Gunaratne and Samarajiva, "Estimating Internet Users."

8. GSMA and A. T. Kearney, "The Mobile Economy 2013" (London, 2013), www.atkearney.com/documents/10192/760890/The_Mobile_Economy_2013.pdf.

9. International Telecommunication Union, "Key 2005-2014 ICT Data."

10. Ingrid Lunden, "Facebook Passes 1B Mobile Users, 200M Messenger Users In Q1," *Techcrunch*, April 23, 2014, http://techcrunch.com/2014/04/23/facebook-passes -1b-mobile-monthly-active-users-in-q1-as-mobile-ads-reach-59-of-all-ad-sales/.

11. Neil Gough, "Chinese Now Prefer Mobile When Going Online," *The New York Times Online*, July 22, 2014, http://sinosphere.blogs.nytimes.com/2014/07/22/ smartphones-surpass-computers-for-internet-use-in-china/; CINIC, *Statistical Report on Internet Development in China* (Beijing, 2014), http://www1.cnnic.cn/IDR/ ReportDownloads/201411/P020141102574314897888.pdf.

12. Ronald Deibert, "Trouble at the Border: China's Internet," *Index on Censorship* 42, no. 2 (2013): 132–135, doi:10.1177/0306422013495334.

13. Statcounter, "Mobile vs. Desktop," *Statcounter Global Stats*, 2013, http:// gs.statcounter.com/. Mary Meeker depicts this crossover in India in Meeker, "2012 Internet Trends @Stanford–Bases."

14. Benedict Evans, "Mobile Is Eating the World," *Andreessen Horowitz*, 2014, http:// a16z.com/2014/10/28/mobile-is-eating-the-world/.

15. International Telecommunication Union, *Measuring the Information Society Report 2014* (Geneva, 2014), http://www.itu.int/en/ITU-D/Statistics/Documents/ publications/mis2014/MIS2014_without_Annex_4.pdf, 4.

16. Ericsson, "Ericsson Mobility Report" (November 2014), 16. Seventy percent of the developing world figure is from the 2013 report. Ericsson, "Ericsson Mobility Report: On the Pulse of the Networked Society" (June 2013), http://www.ericsson .com/res/docs/2013/ericsson-mobility-report-june-2013.pdf.

17. Ericsson, "Ericsson Mobility Report" (November 2014), 16.

18. Ibid.

19. World Bank, *Information and Communications for Development 2012: Maximizing Mobile* (Washington, DC: The World Bank, 2012), doi:10.1596/978-0-8213-8991-1; Tim Kelly and Carlo Maria Rossotto, eds., *Broadband Strategies Handbook* (Washington, DC: The World Bank, 2012); Chris S. Thomas, "Reaching the Third Billion: Arriving at Affordable Broadband to Stimulate Economic Transformation in Emerging Markets," in *The Global Information Technology Report 2012*, ed. Soumitra Dutta

and Beñat Bilbao-Osorio (Geneva: The World Economic Forum and UNESCO, 2012), 79–87; Alliance for Affordable Internet, "The Affordability Report 2013" (Washington, DC, December 10, 2013), http://a4ai.org/wp-content/uploads/2013/12/Affordability-Report-2013-FINAL.pdf.

20. Earl A. Oliver and Srinivasan Keshav, "Design Principles for Opportunistic Communication in Constrained Computing Environments," in *Proceedings of the 2008 ACM Workshop on Wireless Networks and Systems for Developing Regions—WiNS-DR '08* (New York: ACM Press, 2008), 31, doi:10.1145/1410064.1410071.

21. Vikas Sehgal, Kevin Dehoff, and Ganesh Panneer, "The Importance of Frugal Engineering," *Strategy+Business*, no. 59 (May 25, 2010): 20–25; Richard Heeks, "IT Innovation for the Bottom of the Pyramid," *Communications of the ACM* 55, no. 12 (December 1, 2012): 24–27, doi:10.1145/2380656.2380665; Richard T. Watson, K. Niki Kunene, and M. Sirajul Islam, "Frugal Information Systems (IS)," *Information Technology for Development* 19, no. 2 (April 26, 2013): 176–187, doi:10.1080/02681102.2012.714349.

22. The World Bank, *World Development Indicators 2008* (Washington, DC: The World Bank, 2008), http://data.worldbank.org/sites/default/files/wdi08.pdf, p. 4.

23. Ron Amadeo, "Motorola CEO Talks $50 Smartphone, Customizable Screen Sizes," *Ars Technica*, January 22, 2014, http://arstechnica.com/gadgets/2014/01/motorola-ceo-talks-50-smartphone-customizable-screen-sizes/.

24. Leo Mirani, "The Smartphone Companies That Shook up India and China Are Ready to Colonize the World," *Quartz*, October 25, 2013, http://qz.com/139691/the-smartphone-companies-that-shook-up-india-and-china-are-ready-to-colonize-the-world/.

25. Jason Dedrick, Kenneth L. Kraemer, and Greg Linden, "The Distribution of Value in the Mobile Phone Supply Chain," *Telecommunications Policy* 35, no. 6 (July 2011): 505–521, doi:10.1016/j.telpol.2011.04.006.

26. Victor Mulas, "Policies for Mobile Broadband," in *Information and Communications for Development 2012: Maximizing Mobile*, ed. World Bank (Washington, DC: World Bank, 2012), 103–112.

27. Ann K. Armstrong, Joseph J. Mueller, and Tim Syrett, "The Smartphone Royalty Stack: Surveying Royalty Demands for the Components Within Modern Smartphones," *SSRN Electronic Journal* (2014), doi:10.2139/ssrn.2443848.

28. Nehaa Chaudhari, "Methodology: Sub Hundred Dollar Mobile Devices and Competition Law," *CIS Blog*, November 25, 2014, http://cis-india.org/a2k/blogs/methodology-sub-hundred-dollar-mobile-devices-and-competition-law.

29. Martin Kenney and Bryan Pon, "Structuring the Smartphone Industry: Is the Mobile Internet OS Platform the Key?," *Journal of Industry, Competition and Trade* 11, no. 3 (June 7, 2011): 239–261, doi:10.1007/s10842-011-0105-6.

30. IDC, "Worldwide Smartphone Shipments Edge Past 300 Million Units in the Second Quarter; Android and iOS Devices Account for 96% of the Global Market, According to IDC," Press Release, August 14, 2014, http://www.idc.com/getdoc.jsp ?containerId=prUS25037214.

31. Dan Seifert, "With Android One, Google Is Poised to Own the Entire World," *The Verge*, June 26, 2014, http://www.theverge.com/2014/6/26/5845562/android -one-google-the-next-billion.

32. BBC, "Mozilla Plans '$25 Smartphone' for Emerging Markets," *BBC News Online*, February 23, 2014, http://www.bbc.com/news/technology-26316265.

33. James Goodman, "Return to Vendor: How Second-Hand Mobile Phones Improve Access to Telephone Services" (London: Forum for the Future, 2004).

34. Nimmi Rangaswamy and Sumitra Nair, "The Mobile Phone Store Ecology in a Mumbai Slum Community: Hybrid Networks for Enterprise," *Information Technologies & International Development* 6, no. 3 (2010): 51–65.

35. S. P. Ketkar, "Innovative Scheme with BPL Connections Routed via BSNL Will Promote Financial Inclusion," *The Economic Times*, September 13, 2012, http:// articles.economictimes.indiatimes.com/2012-09-13/news/33817058_1_bpl-families-wireless-subscribers-scheme; BBC, "Nigeria: ANPP Anger over Free Phone Plan for Farmers," *BBC World News Online*, January 8, 2013, http://www.bbc.com/news/ world-africa-20947686.

36. Jeffrey James, *Digital Interactions in Developing Countries: An Economic Perspective* (New York: Routledge, 2013).

37. Jenna Burrell, "Evaluating Shared Access: Social Equality and the Circulation of Mobile Phones in Rural Uganda," *Journal of Computer-Mediated Communication* 15, no. 2 (January 2010): 230–250, doi:10.1111/j.1083-6101.2010.01518.x; Lorenzo Dalvit, "Why Care about Sharing? Shared Phones and Shared Networks in Rural Areas," *Rhodes Journalism Review* 34 (August 2014): 81–85.

38. Molly Wright Steenson and Jonathan Donner, "Beyond the Personal and Private: Modes of Mobile Phone Sharing in Urban India," in *The Reconstruction of Space and Time: Mobile Communication Practices*, ed. Scott W. Campbell and Richard Ling (Piscataway, NJ: Transaction Books, 2007), 231–250.

39. Salahuddin Aminuzzaman, Harald Baldersheim, and Ishtiaq Jamil, "Talking Back! Empowerment and Mobile Phones in Rural Bangladesh: A Study of the Village Phone Scheme of Grameen Bank," *Contemporary South Asia* 12, no. 3 (January 1, 2003): 327–348, doi:10.1080/0958493032000175879.

40. Bjorn Van Campenhout, *Is There an App for That? The Impact of Community Knowledge Workers in Uganda*, International Food Policy Research Institute Discussion Paper 01316, 2013, http://www.ifpri.org/sites/default/files/publications/ ifpridp01316.pdf.

41. Sindhura Chava, Rachid Ennaji, Jay Chen, and Lakshminarayanan Subramanian, "Cost-Aware Mobile Web Browsing," *IEEE Pervasive Computing* 11, no. 3 (March 1, 2012): 34–42, doi:10.1109/MPRV.2012.19.

42. Cisco, "Cisco Visual Networking Index: Global Mobile Data Traffic Forecast Update 2014–2019 White Paper," *Cisco Whitepapers* (San Jose, CA, 2015), http://www.cisco.com/c/en/us/solutions/collateral/service-provider/visual-networking-index-vni/white_paper_c11-520862.html.

43. W3C, "Mobile Web Best Practices 1.0," July 29, 2008, http://www.w3.org/TR/mobile-bp/.

44. W3C, "Mobile Web Application Best Practices," December 14, 2010, http://www.w3.org/TR/mwabp/.

45. Susana Ferreira and Will Connos, "In These Countries, BlackBerry Is Still King—Of Pop Culture," *The Wall Street Journal*, September 11, 2012, http://online.wsj.com/article/SB10000872396390444082904577605552824161264.html-articleTabs=article.

46. Reza H. Namavar, "BlackBerry's Secret Weapon," *Seeking Alpha*, August 21, 2012, http://seekingalpha.com/article/819291-blackberry-s-secret-weapon.

47. Cade Metz, "Why Google Is Following Facebook's WhatsApp Gambit," *Wired.com*, March 10, 2014, http://www.wired.com/2014/10/google-following-facebooks-19-billion-whatsapp-gambit/.

48. Chava et al., "Cost-Aware Mobile Web Browsing."

49. Timo Lehtonen, Said Benamar, Vesa Laamanen, Ilkka Luoma, Olli Ruotsalainen, Jaako Salonen, and Tommi Mikkonen, "Towards User-Friendly Mobile Browsing," in *Proc AAA-IDEA '06* (New York: ACM Press, 2006), 6, doi:10.1145/1190183.1190190.

50. Opera Software, "The News Hungry Nordic Countries (January 2014)," *The State of the Mobile Web* (Oslo, January 2014), http://www.operasoftware.com/smw/2014-01.

51. Frederic Lardinois, "Opera Opens Pre-Registration for Its Data-Savings Android App," *Techcrunch*, February 8, 2014, http://techcrunch.com/2014/02/18/opera-opens-pre-registration-for-its-data-savings-android-app/.

52. Rachel Metz, "Free App Makes 'Dumb' Phones Smarter and Faster," *Technology Review*, August 24, 2012, http://mashable.com/2012/08/24/binu-java-app/.

53. Matthew Kirk, Steve Bratt, and Diane Coyle, *Making Broadband Accessible for All* (London: Vodafone Group, 2011); Mulas, "Policies for Mobile Broadband"; Kelly and Rossotto, *Broadband Strategies Handbook*; Thomas, "Reaching the Third Billion."

54. Martin Cave and Windfred Mfuh, "Rethinking Mobile Regulation for the Data Age," in Kirk, Bratt, and Coyle, *Making Broadband Accessible for All*, 41–48.

55. Diane Coyle and Howard Williams, "Overview," in Kirk, Bratt, and Cole, *Making Broadband Accessible for All*, 3–11.

56. Philip N. Howard and Nimah Mazaheri, "Telecommunications Reform, Internet Use and Mobile Phone Adoption in the Developing World," *World Development* 37, no. 7 (July 2009): 1159–1169, doi:10.1016/j.worlddev.2008.12.005.

57. Rohan Samarajiva, "Leveraging the Budget Telecom Network Business Model to Bring Broadband to the People," *Information Technologies & International Development* 6, no. SE (2010): 93–97.

58. Piet Buys, Susmita Dasgupta, Timothy S. Thomas, and David Wheeler, "Determinants of a Digital Divide in Sub-Saharan Africa: A Spatial Econometric Analysis of Cell Phone Coverage," *World Development* 37, no. 9 (September 2009): 1494–1505, doi:10.1016/j.worlddev.2009.01.011; Kirk, Bratt, and Coyle, *Making Broadband Accessible for All*.

59. Kas Kalba, "The Adoption of Mobile Phones in Emerging Markets: Global Diffusion and the Rural Challenge," *International Journal of Communication* 2 (2008): 631–661.

60. Andrew Dymond and Sonja Oestmann, *Information and Communication Technologies (ICTs), Poverty Alleviation, and Universal Access Policies* (Nairobi: African Technology Policy Studies Network, 2002).

61. Ibid., 8.

62. Benjamin M. Compaine and Mitchell J. Weinraub, "Universal Access to Online Services: An Examination of the Issue," *Telecommunications Policy* 21, no. 1 (February 1997): 15–33, doi:10.1016/S0308-5961(96)00062-6.

63. Buys et al., "Determinants of a Digital Divide"; Mulas, "Policies for Mobile Broadband," p. 106.

64. Mulas, "Policies for Mobile Broadband."

65. Russell Southwood, "Africa's Future Data Architecture Beginning to Fall into Place—Exchange Points and Data Centres," *Balancing Act Africa*, October 5, 2012, http://www.balancingact-africa.com/news/en/issue-no-625/top-story/africa-s-future -data/en.

66. Marshini Chetty, Srikanth Sundaresan, Sachit Muckaden, Nick Feamster, and Enrico Calandro, "Measuring Broadband Performance in South Africa," in *Proceedings of the 4th Annual Symposium on Computing for Development—ACM DEV-4 '13* (New York: ACM Press, 2013), 1–10, doi:10.1145/2537052.2537053.

67. Marshini Chetty, Richard Banks, A. J. Bernheim Brush, Jonathan Donner, and Rebecca E. Grinter, "'You're Capped!' Understanding the Effects of Bandwidth Caps on Broadband Use in the Home," in *Proceedings of CHI 2012, May 5–10, 2010, Austin, TX* (New York: ACM, 2012), 3021–3030, doi:10.1145/2207676.2208714.

68. Christopher T. Marsden, "Network Neutrality: A Research Guide," in *Research Handbook on Governance of the Internet.*, ed. Ian Brown (Cheltenham, UK: Edward Elgar, 2013), 419–444, doi:10.4337/9781849805049.00026.

69. Fenwick McKelvey, "Ends and Ways: The Algorithmic Politics of Network Neutrality," *Global Media Journal—Canadian Edition* 3, no. 1 (2010): 51–73.

70. Thomas W. Hazlett, "Spectrum Policy and Competition in Mobile Services," in Kirk, Bratt, and Coyle, *Making Broadband Accessible for All*, 31–40.

71. Mulas, "Policies for Mobile Broadband," 105.

72. Mobile World Live, "Nokia and Bharti Eye Next Billion Internet Connections," *Mobile World Live*, February 26, 2013, http://www.mobileworldlive.com/nokia-and-bharti-eye-next-billion-internet-connections; Russell Southwood, "Google and Microsoft Pitch TV White Spaces as a Way to Wring out More Spectrum and Beat the Digital Divide in Rural Areas," *Balancing Act Africa*, May 31, 2013, http://www.balancingact-africa.com/news/en/issue-no-657/top-story/google-and-microsoft/en.

73. Hsien-Tang Ko, Chien-Hsun Lee, Jia-Huei Lin, Kay Chung, and Nan-Shiun Chu, "Television White Spaces: Learning from Cases of Recent Trials," *International Journal of Digital Television* 5, no. 2 (June 1, 2014): 149–167, doi:10.1386/jdtv.5.2.149_1.

74. Richard Heeks, "Mobiles for Impoverishment?," *ICTs for Development Blog*, December 27, 2008, http://ict4dblog.wordpress.com/2008/12/27/mobiles-for-impoverishment/. Also see description of right2know in Walton and Leukes, "Prepaid Social Media and Mobile Discourse in South Africa," *Journal of African Media Studies* 5, no. 2 (2013): 149–167, doi:10.1386/jams.5.2.149_1.

75. The Broadband Commission, *The State of Broadband 2014: Broadband for All* (Geneva, Switzerland, 2014), http://www.broadbandcommission.org/Documents/reports/bb-annualreport2014.pdf, 39.

76. Susan P. Wyche, Sarita Yardi Schoenebeck, and Andrea Forte, "'Facebook Is a Luxury,'" in *Proceedings of the 2013 Conference on Computer Supported Cooperative Work—CSCW '13* (New York: ACM Press, 2013), 33–44, doi:10.1145/2441776.2441783.

77. Jan Chipchase, "Connectivity Is Not Binary, the Network Is Never Neutral," *Medium*, September 2, 2014, https://medium.com/todays-office/connectivity-is-not-binary-the-network-is-never-neutral-4620b2a26746.

78. Ethan Zuckerman, "Decentralizing the Mobile Phone: A Second ICT4D Revolution?," *Information Technologies & International Development* 6 (2010): 99–103.

79. Vanessa Daly, "Wi-Fi Grows to over 50% of Mobile Web Connections," *Bango.com Blog*, February 2, 2011, http://news.bango.com/2011/02/02/wi-fi-grows-to-over-50-percent/; Mobidia, "Managed Wi-Fi Hotspot Usage," *Understanding Mobile Data*

Blog, February 21, 2013, http://mobidia.blogspot.com/2013/02/managed-wi-fi-hotspot-usage-over-past.html; Cisco, "Cisco Visual Networking Index."

80. Walton and Leukes, "Prepaid Social Media and Mobile Discourse in South Africa."

81. Steve Song, "Unpacking Our Mobile Broadband Future," *Many Possibilities*, October 15, 2012, http://manypossibilities.net/2012/10/unpacking-our-mobile-broadband-future/.

82. Aaron Mason, "Inveneo Enhancing Educational Offerings and Spinning Off a Broadband Startup," *Inveneo.org* (blog), September 12, 2013, http://www.inveneo.org/2013/09/inveneo-enhancing-educational-offerings-and-spinning-off-a-broadband-startup/.

83. Steve Song, "Africa's LTE Future," *Many Possibilities*, January 17, 2014, http://manypossibilities.net/2014/01/africas-lte-future/.

84. Kevin Fitchard, "Exclusive: Airtel Bets Big on Wi-Fi across Africa as It Looks for 3G Substitutes," *GigaOm*, February 21, 2013, http://gigaom.com/2013/02/21/exclusive-airtel-bets-big-on-wi-fi-across-africa-as-it-looks-for-3g-substitutes/.

85. RNW Africa Desk, "Building a Grassroots African Internet," *Radio Netherlands Worldwide*, November 29, 2013, http://www.rnw.nl/africa/article/building-a-grassroots-african-internet.

86. David L. Johnson, Elizabeth M. Belding, and Gertjan van Stam, "Network Traffic Locality in a Rural African Village," in *Proceedings of the Fifth International Conference on Information and Communication Technologies and Development—ICTD '12* (New York: ACM Press, 2012), 268–277, doi:10.1145/2160673.2160707.

87. Hernán Galperin and François Bar, "The Microtelco Opportunity: Evidence from Latin America," *Information Technologies and International Development* 3, no. 2 (2006): 73–86; Abhinav Anand, Veljko Pejovic, Elizabeth M. Belding, and David L. Johnson, "VillageCell: Cost Effective Cellular Connectivity in Rural Areas," in *Proceedings of the Fifth International Conference on Information and Communication Technologies and Development—ICTD '12* (New York: ACM Press, 2012), 180–189, doi:10.1145/2160673.2160698.

88. Kurtis Heimerl, Shaddi Hasan, Kashif Ali, Eric Brewer, and Tapan Parikh, "Local, Sustainable, Small-Scale Cellular Networks," in *Proceedings of the Sixth International Conference on Information and Communication Technologies and Development Full Papers—ICTD '13*, vol. 1 (New York: ACM Press, 2013), 2–12, doi:10.1145/2516604.2516616.

89. Russell Southwood, "Micro Networks May Offer a Way to Get 4G out of Africa's Cities and into Towns and Villages," *Balancing Act Africa*, December 12, 2014, http://www.balancingact-africa.com/news/en/issue-no-737/top-story/micro-networks-may-o/en.

90. Veljko Pejovic, David L. Johnson, Mariya Zheleva, Elizabeth M. Belding, and Lisa Parks, "The Bandwidth Divide: Obstacles to Efficient Broadband Adoption in Rural Sub-Saharan Africa," *International Journal of Communication* 6 (2012): 2467–2491.

91. Alistair Barr and Andy Pasztor, "Google Invests in Satellites to Spread Internet Access," *The Wall Street Journal*, June 1, 2014, http://www.wsj.com/articles/google -invests-in-satellites-to-spread-internet-access-1401666287.

92. Michael Kassner, "Outernet and O3b: Making Internet Access Available to All Using Satellite Technology," *Techrepublic.com*, March 26, 2014, http://www .techrepublic.com/article/outernet-and-o3b-making-internet-access-available-to-all -using-satellite-technology/.

93. Megan Geuss, "Google, Fidelity Invest $1 Billion in SpaceX and Satellite Internet Plan," *Ars Technica*, January 20, 2015, http://arstechnica.com/business/2015/01/ google-might-pour-money-into-spacex-really-wants-satellite-internet/.

94. Michael Kassner, "Outernet and O3b: Making Internet Access Available to All Using Satellite Technology"; Scott Burleigh, "Nanosatellites for Universal Network Access," in *Proceedings of the 2013 ACM MobiCom Workshop on Lowest Cost Denomina-tor Networking for Universal Access—LCDNet '13* (New York: ACM Press, 2013), 33–34, doi:10.1145/2502880.2502896.

95. Steven Levy, "The Untold Story of Google's Quest to Bring the Internet Everywhere—By Balloon," *Wired.com*, August 13, 2013, http://www.wired.com/ 2013/08/googlex-project-loon/.

96. Vindu Goel, "A New Facebook Lab Is Intent on Delivering Internet Access by Drone," *New York Times*, March 27, 2014, http://www.nytimes.com/2014/03/28/ technology/a-new-facebook-lab-is-intent-on-delivering-internet-access-by-drone .html.

Chapter 3

1. GSMA and A. T. Kearney, "The Mobile Economy 2013" (London 2013), www.atkearney.com/documents/10192/760890/The_Mobile_Economy_2013.pdf.

2. Roshanthi Lucas Gunaratne and Rohan Samarajiva, "Estimating Internet Users: An Evidence-Based Alternative in the Absence of Survey Data," *Info* 15, no. 5 (2013): 20–33, doi:10.1108/info-05-2013-0023.

3. Search performed January 2015 via http://scholar.google.com/scholar?q=allintitl e%3A+"the+mobile+internet"&btnG=&hl=en&as_sdt=0,5.

4. Gerard Goggin, "Driving the Internet: Mobile Internets, Cars, and the Social," *Future Internet* 4, no. 4 (March 20, 2012): 306–321, doi:10.3390/fi4010306.

5. Paul Levinson, *Cellphone: The Story of the World's Most Mobile Medium* (New York: Palgrave MacMillan, 2004).

6. Leslie Horn, "First Portable Computer Debuted 30 Years Ago," *PC Mag Digital Edition*, April 4, 2011, http://www.pcmag.com/article2/0,2817,2383022,00.asp.

7. Jason Fell, "A True Transformer: Asus Announces a Laptop-Tablet That Runs Android and Windows," *Entrepreneur*, January 8, 2014, http://www.entrepreneur.com/article/230703.

8. Kevin Gallo, "A First Look at the Windows 10 Universal App Platform," *Microsoft Windows Blog*, March 2, 2015, http://blogs.windows.com/buildingapps/2015/03/02/a-first-look-at-the-windows-10-universal-app-platform/.

9. Jason Perlow, "Ubuntu Edge: A Grand Experiment for the Future of Computing Does Not Constitute a Failure," *ZDNet*, August 22, 2013, http://www.zdnet.com/ubuntu-edge-a-grand-experiment-for-the-future-of-computing-does-not-constitute-a-failure-7000019762/.

10. Marshini Chetty, Srikanth Sundaresan, Sachit Muckaden, Nick Feamster, and Enrico Calandro, "Measuring Broadband Performance in South Africa," in *Proceedings of the 4th Annual Symposium on Computing for Development—ACM DEV-4 '13* (New York: ACM Press, 2013), 1–10, doi:10.1145/2537052.2537053.

11. Kevin Hannam, Mimi Sheller, and John Urry, "Editorial: Mobilities, Immobilities and Moorings," *Mobilities* 1, no. 1 (March 2006): 1–22, doi:10.1080/17450100500489189; Klaus Bruhn Jensen, "What's Mobile in Mobile Communication?," *Mobile Media & Communication* 1, no. 1 (January 1, 2013): 26–31, doi:10.1177/2050157912459493.

12. Gerard Goggin, "Mobile Video: Spreading Stories with Mobile Media," in *Routledge Companion to Mobile Media*, ed. Gerard Goggin and Larissa Hjorth (New York: Routledge, 2014), 146–156.

13. Google, "Mobile Search Moments," *Google Think Insights*, March 2013, http://www.google.com/think/research-studies/creating-moments-that-matter.html.

14. Nicola Green, "On the Move: Technology, Mobility, and the Mediation of Social Time and Space," *The Information Society* 18, no. 4 (July 2002): 281–292, doi:10.1080/01972240290075129.

15. James E. Katz and Mark Aakhus, "Conclusion: Making Meaning of Mobiles: A Theory of Apparatgeist," in *Perpetual Contact: Mobile Communication, Private Talk, and Public Performance*, ed. James E. Katz and Mark Aakhus (Cambridge, UK: Cambridge University Press, 2002), 301–318; Raul Pertierra, "Mobile Phones, Identity and Discursive Intimacy," *Human Technology* 1, no. 1 (2005): 23–44.

16. Richard Ling, *Taken for Grantedness* (Cambridge, MA: MIT Press, 2012).

17. Finn Trosby, "SMS, the Strange Duckling of GSM," *Telektronikk* 3 (2004): 187–194.

18. Geoff Huston, "TCP in a Wireless World," *IEEE Internet Computing* 5, no. 2 (2001): 82–84, doi:10.1109/4236.914651.

19. Kenichi Ishii, "Internet Use via Mobile Phone in Japan," *Telecommunications Policy* 28, no. 1 (February 2004): 43–58, doi:10.1016/j.telpol.2003.07.001.

20. Joel West and Michael Mace, "Browsing as the Killer App: Explaining the Rapid Success of Apple's iPhone," *Telecommunications Policy* 34, no. 5–6 (June 2010): 270–286, doi:10.1016/j.telpol.2009.12.002.

21. Jerry Watkins, Larissa Hjorth, and Ilpo Koskinen, "Wising up: Revising Mobile Media in an Age of Smartphones," *Continuum* 26, no. 5 (October 2012): 665–668, doi:10.1080/10304312.2012.706456. P. 665.

22. Andrew Grantham and George Tsekouras, "Information Society: Wireless ICTs' Transformative Potential," *Futures* 36, no. 3 (April 2004): 359–377, doi:10.1016/S0016-3287(03)00066-1; Ben Agger, "iTime: Labor and Life in a Smartphone Era," *Time & Society* 20, no. 1 (April 20, 2011): 119–136, doi:10.1177/0961463X10380730.

23. Karissa Bell, "Google Play Now Has More Apps than Apple's App Store, Report Says," *Mashable*, January 15, 2015, http://mashable.com/2015/01/15/google-play-more-apps-than-ios/.

24. Troes F. Bertel and Gitte Stald, "From SMS to SNS: The Use of the Internet on the Mobile Phone among Young Danes," in *Mobile Media Practices, Presence and Politics: The Challenge of Being Seamlessly Mobile*, ed. Kathleen M. Cumiskey and Larissa Hjorth (New York: Routledge, 2013), 198–213.

25. Ran Makavy, "Feature Phone Milestone: Facebook for Every Phone Reaches 100 Million," *Facebook Newsroom*, July 21, 2013, http://newsroom.fb.com/News/663/Feature-Phone-Milestone-Facebook-for-Every-Phone-Reaches-100-Million.

26. Vikas Sehgal, Kevin Dehoff, and Ganesh Panneer, "The Importance of Frugal Engineering," *Strategy+Business*, no. 59 (May 25, 2010): 20–25.

27. Yuri Dikhanov, "Trends in Global Income Distribution, 1970–2000, and Scenarios for 2015," *Human Development Occasional Papers (1992–2007)* (Geneva: UNDP, September 2005), http://hdr.undp.org/en/reports/global/hdr2005/papers/HDR2005_Dikhanov_Yuri_8.pdf.

28. Steve Costello, "Smartphone Shipments Set to Surpass Feature Phones," *Mobile World Live*, March 5, 2013, http://www.mobileworldlive.com/smartphone-shipments-set-to-surpass-feature-phones; Ericsson, "Ericsson Mobility Report: On the Pulse of the Networked Society 2013" (June 2013), http://www.ericsson.com/res/docs/2013/ericsson-mobility-report-june-2013.pdf.

29. Ericsson, "Ericsson Mobility Report: On the Pulse of the Networked Society 2013."

30. Mary Meeker, "Internet Trends 2014: Code Conference," *KPCB News*, May 28, 2014, 73, http://kpcbweb2.s3.amazonaws.com/files/85/Internet_Trends _2014_vFINAL_-_05_28_14-_PDF.pdf?1401286773.

31. Balancing Act Africa, "Smartphones vs Feature Phones—The Handset War That Will Shape the Coming Transition to All-Data Services," April 5, 2013, http:// www.balancingact-africa.com/news/en/issue-no-649/top-story/smartphones-vs -featu/en; GSMA Mobile for Development Intelligence, *Scaling Mobile for Development: Harness the Opportunity* (London, August 2013), https://gsmaintelligence.com/ research/?file=130828-scaling-mobile.pdf.

32. Jeremy Ford, "$80 Android Phone Sells Like Hotcakes in Kenya, the World Next?," *Singularity Hub*, July 16, 2011, http://singularityhub.com/2011/08/16/80 -android-phone-sells-like-hotcakes-in-kenya-the-world-next/.

33. Natasha Lomas, "$50 Android Smartphones Are Disrupting Africa Much Faster Than You Think, Says Wikipedia's Jimmy Wales," *Techcrunch*, December 10, 2013, http://techcrunch.com/2012/12/10/50-android-smartphones-are-disrupting-africa -much-faster-than-you-think-says-wikipedias-jimmy-wales/.

34. Shunal Doke, "Nokia Has Launched One of Its Latest Phones from the Asha Line in India," *Tech2*, October 22, 2012, http://tech.firstpost.com/news-analysis/nokia -asha-308-launched-in-india-for-rs-5865-49924.html.

35. Ron Amadeo, "Testing a $35 Firefox OS Phone—How Bad Could It Be?," *Ars Technica*, October 7, 2014, http://arstechnica.com/gadgets/2014/10/testing-a-35 -firefox-os-phone-how-bad-could-it-be/.

36. Steve Costello, "Android Heading to 1B User Milestone," *Mobile World Live*, April 17, 2013, http://www.mobileworldlive.com/android-heading-to-1b-user -milestone.

37. Samita Thapa, "Five Smartphones for under $50 USD," *TechChange*, July 30, 2014, http://techchange.org/2014/07/30/cheap-smartphones-under-50 -dollars/.

38. OAfrica, "African Smartphone Usage Driven by Savvy Young South African, Kenyan, and Nigerian Consumers," *OAfrica Blog*, December 7, 2013, http:// www.oafrica.com/mobile/african-smartphone-usage-driven-by-savvy-young-south -african-kenyan-and-nigerian-consumers/.

39. Ericsson, "Ericsson Mobility Report: On the Pulse of the Networked Society 2014" (Stockholm, November 2014), http://www.ericsson.com/res/docs/2014/ ericsson-mobility-report-november-2014.pdf.

40. Natasha Lomas, "Talking on Tablets on the Rise in Asia," *Techcrunch*, August 20, 2014, http://techcrunch.com/2014/08/20/im-on-the-pad-phone/; Kevin C. Tofel, "Why Are Phones Continuing to Get Bigger? (Hint: Think Post-PC)," *GigaOm*, January 2, 2014, http://gigaom.com/2014/01/02/why-are-phones-continuing-to-get -bigger-hint-think-post-pc/.

41. Luke van Hooft, "Building Next Generation Broadband Networks in Emerging Markets," *Making Broadband Accessible for All—Policy Papers Series #12* (London: Vodafone Group, May 2011), http://www.vodafone.com/content/dam/vodafone/ about/public_policy/policy_papers/public_policy_series_12.pdf.

42. Claudio Feijóo, "Next Generation Mobile Networks and Technologies: Impact on Mobile Media," in *Routledge Companion to Mobile Media*, ed. Gerard Goggin and Larissa Hjorth (New York: Routledge, 2014), 81–91; Indra de Lanerolle, Alison Gill-wald, Christoph Stork, and Enrico Calandro, "Let Them Eat Movies: (How) Will Next-Generation Broadband Diffuse Through Africa?," in *Broadband as a Video Plat-form: Strategies for Africa*, ed. Judith O'Neill, Eli M. Noam, and Darcy Gerbarg, The Economics of Information, Communication, and Entertainment (Geneva: Springer International Publishing, 2014), 15–40, doi:10.1007/978-3-319-03617-5; Eli Noam, "Let Them Eat Cellphones: Why Mobile Wireless Is No Solution for Broadband," *Journal of Information Policy* 1 (2011): np.

43. Ericsson, "Ericsson Mobility Report: On the Pulse of the Networked Society 2014,"24.

44. Cisco, "Cisco Visual Networking Index: Global Mobile Data Traffic Forecast Update, 2013–2018," (San Jose CA: 2014), https://web.archive.org/web/ 20140302213838/http://www.cisco.com/c/en/us/solutions/collateral/service -provider/visual-networking-index-vni/white_paper_c11-520862.html?.

45. Ericsson, "Ericsson Mobility Report: On the Pulse of the Networked Society 2014,"22.

46. Ericsson, "Ericsson Mobility Report: On the Pulse of the Networked Society 2013"; Cisco, "Cisco Visual Networking Index 2013–2018."

47. Ericsson, "Ericsson Mobility Report: On the Pulse of the Networked Society 2014," 16.

48. Helen Nyambura-Mwaura and Simon Akam, "Telecoms Boom Leaves Rural Africa Behind," *Reuters*, January 31, 2013, http://www.reuters.com/article/2013/ 01/31/us-africa-telecoms-idUSBRE90U0MK20130131; GSMA Mobile for Develop-ment Intelligence, *Scaling Mobile for Development*.

49. Trosby, "SMS, the Strange Duckling of GSM."

50. Richard Ling and Jonathan Donner, *Mobile Communication* (Cambridge, UK: Polity, 2009).

51. Neha Dharia, "Global SMS Revenues Will Decline after 2016," *Ovum Analyst Opinions*, November 11, 2013, http://www.ovum.com/global-sms-revenues-will -decline-after-2016/.

52. Naomi S. Baron, *Always On: Language in an Online and Mobile World* (New York: Oxford University Press, 2008).

53. Gerard Goggin and Christina Spurgeon, "Premium Rate Culture: The New Business of Mobile Interactivity," *New Media & Society* 9, no. 5 (October 1, 2007): 753–770, doi:10.1177/1461444807080340.

54. Jay Chen, Lakshmi Subramanian, and Eric Brewer, "SMS-Based Web Search for Low-End Mobile Devices," in *Proceedings of the Sixteenth Annual International Conference on Mobile Computing and Networking—MobiCom '10* (New York: ACM Press, 2010), 125–136, doi:10.1145/1859995.1860011.

55. Google, "Google SMS Applications," *Google Mobile*, February 8, 2013, http://www.google.com/mobile/sms/.

56. Kul Wadhwa and Howie Fung, "Converting Western Internet to Indigenous Internet: Lessons from Wikipedia," *Innovations: Technology, Governance, Globalization* 9, no. 3–4 (July 2014): 132–141, doi:10.1162/inov_a_00224.

57. SmartSMS, "About Smart SMS," 2015, http://smartsms.net/about/.

58. Ken Banks, Sean Martin McDonald, and Florence Scialom, "Mobile Technology and the Last Mile: 'Reluctant Innovation' and FrontlineSMS," *Innovations: Technology, Governance, Globalization* 6, no. 1 (January 18, 2011): 7–12, doi:10.1162/INOV_a_00055.

59. Ory Okolloh, "Ushahidi, or 'Testimony': Web 2.0 Tools for Crowdsourcing Crisis Information," *Participatory Learning and Action* 59, no. 1 (2009): 65–70.

60. Staff writer, "MTN SurferLite—Low Cost Internet for Feature Phones," *MyBroadband*, December 11, 2013, http://mybroadband.co.za/news/cellular/93441-mtn-surferlite-low-cost-internet-for-feature-phones.html.

61. Christopher Mims, "Facebook's Plan to Find Its Next Billion Users: Convince Them the Internet and Facebook Are the Same," *Quartz*, September 24, 2012, http://qz.com/5180/facebooks-plan-to-find-its-next-billion-users-convince-them-the -internet-and-facebook-are-the-same/.

62. Mediawiki.org, "Wikipedia over SMS & USSD/status," *Mediawiki.org*, March 3, 2014, http://www.mediawiki.org/wiki/Wikipedia_over_SMS_&_USSD/status. See also Merryl Ford and Adele Botha, "MobilED–An Accessible Mobile Learning Platform for Africa," in *IST-Africa 2007 Conference Proceedings*, ed. Paul Cunningham and Miriam Cunningham (IIMC International Information Management Corporation, 2007), 9–11.

63. Aditya Vashistha and William Thies, "IVR Junction: Building Scalable and Distributed Voice Forums in the Developing World," 6th USENIX/ACM Workshop on Networked Systems for Developing Regions, June 15, 2012, https://www.usenix.org/system/files/conference/nsdr12/nsdr12-final4.pdf.

64. Arun Kumar, Sheetal K. Agarwal, and Priyanka Manwani, "The Spoken Web Application Framework," in *Proceedings of the 2010 International Cross Disciplinary Conference on Web Accessibility (W4A)—W4A '10* (New York: ACM Press, 2010), np, doi:10.1145/1805986.1805990.

65. Neil Patel, Deepti Chittamuru, Anupam Jain, Paresh Dave, and Tapan S. Parikh, "Avaaj Otalo: A Field Study of an Interactive Voice Forum for Small Farmers in Rural India," in *Proceedings of the 28th International Conference on Human Factors in Computing Systems—CHI '10* (New York: ACM Press, 2010), 733–742, doi:10.1145/1753326.1753434.

66. Arun Kumar, Nitendra Rajput, Sheetal Agarwal, Dipanjan Chakraborty, and Amit Anit Nanavati, "Organizing the Unorganized—Employing IT to Empower the Under-Privileged," in *Proceedings of the 17th International Conference on World Wide Web—WWW '08* (New York: ACM Press, 2008), 935–944, doi:10.1145/1367497.1367623.

67. Sara Chamberlain, "A Mobile Guide Toward Better Health: How Mobile Kunji Is Improving Birth Outcomes in Bihar, India," *Innovations: Technology, Governance, Globalization* 9, no. 3–4 (July 2014): 43–52, doi:10.1162/inov_a_00215.

68. Juhee Kang and Moutusy Maity, "Texting among the Bottom of the Pyramid: Facilitators and Barriers to SMS Use among the Low-Income Mobile Users in Asia," Draft Report for LIRNEasia (Colombo, Sri Lanka, August 2012), http://lirneasia.net/wp-content/uploads/2010/07/Texting-among-the-Bottom-of-the-Pyramid-Facilitators-and-Barriers-to-SMS-Use-among-the-Low-income-Mobile-Users-in-Asia.pdf.

69. Preeti Mudliar, Jonathan Donner, and William Thies, "Emergent Practices around CGNet Swara: A Voice Forum for Citizen Journalism in Rural India," *Information Technologies & International Development* 9, no. 2 (2013): 65–79.

70. Luke Goode, "Social News, Citizen Journalism and Democracy," *New Media & Society* 11, no. 8 (November 24, 2009): 1287–1305, doi:10.1177/1461444809341393.

71. Nithya Sambasivan, Edward Cutrell, Kentaro Toyama, and Bonnie A. Nardi, "Intermediated Technology Use in Developing Communities," in *Proceedings of the 28th International Conference on Human Factors in Computing Systems—CHI '10* (New York: ACM Press, 2010), 2583–2592, doi:10.1145/1753326.1753718; Judith Mariscal and Angelica Martinez, *The Informational Life of the Marginalized: A Study of Digital Access in Three Mexican Towns* (Lima, Peru: DRSI, 2014).

72. Chamberlain, "A Mobile Guide Toward Better Health."

73. Jessica Osborn, "MOTECH," in *mHealth in Practice: Mobile Technology for Health Promotion in the Developing World*, ed. Jonathan Donner and Patricia Mechael (London: Bloomsbury Academic, 2012), 98–116.

74. GSMA and A. T. Kearney, *The Mobile Economy 2013* (London, 2013), http://www.gsmamobileeconomy.com/GSMA%20Mobile%20Economy%202013.pdf; Ericsson, "Ericsson Mobility Report: On the Pulse of the Networked Society 2013"; GSMA Mobile for Development Intelligence, *Scaling Mobile for Development*.

75. Gerard Goggin, "Driving the Internet: Mobile Internets, Cars, and the Social," *Future Internet* 4, no. 4 (March 20, 2012): 306–321, doi:10.3390/fi4010306, p. 307.

76. Gerard Goggin, "Global Internets," in *The Handbook of Global Media Research*, ed. Ingrid Volkmer (Oxford, UK: Wiley-Blackwell, 2012), 352–364, doi:10.1002/9781118255278.ch20. See also Gary Marsden, "What Is the Mobile Internet?" *Interactions* 14, no. 6 (November 2007): 24–25, doi:10.1145/1300655.1300672.

77. Brenda Danet and Susan C. Herring, eds., *The Multilingual Internet* (Oxford, UK: Oxford University Press, 2007); Katy E. Pearce and Ronald E. Rice, "The Language Divide—The Persistence of English Proficiency as a Gateway to the Internet: The Cases of Armenia, Azerbaijan, and Georgia," *International Journal of Communication* 8 (2014): 2834–2859; Iris Orriss, "The Internet's Language Barrier," *Innovations: Technology, Governance, Globalization* 9, no. 3–4 (July 2014): 123–126, doi:10.1162/inov_a_00223.

78. Daniel Miller and Don Slater, *The Internet: An Ethnographic Approach* (Oxford, UK: Berg, 2000).

79. Goggin, "Global Internets."

80. Elizabeth Davison and Shelia R. Cotten, "Connection Discrepancies: Unmasking Further Layers of the Digital Divide," *First Monday* 8, no. 3 (March 3, 2003), doi:10.5210/fm.v8i3.1039.

81. Arthur Goldstuck, "The Mobile Internet Pinned Down," *World Wide Worx Reports* (Johannesburg, South Africa, May 27, 2010), http://www.worldwideworx.com/the-mobile-internet-pinned-down/; Rohan Samarajiva, "Explaining Why 32 Million Indonesian Facebook Users Are Not Counted by Their Government as Internet Users," *LIRNEasia Blog*, September 1, 2014, http://lirneasia.net/2014/09/explaining-why-32-million-indonesians-who-use-facebook-are-not-counted-by-their-government-as-internet-users/; Jonathan Donner and Cecile Bezuidenhoudt, "Due to Mobile Data, Survey Questions about Internet Use Should No Longer Implicitly Favor the PC," in *Mobile Media Practices, Presence and Politics: The Challenge of Being Seamlessly Mobile*, ed. Katie Cumiskey and Larissa Hjorth (New York: Routledge, 2013), 83–97.

82. Jonathan Donner and Shikoh Gitau, "New Paths: Exploring Mobile-Centric Internet Use in South Africa," "Mobile 2.0: Beyond Voice?" Pre-Conference Workshop at the International Communication Association (ICA) (Chicago, April 2009), http://lirneasia.net/wp-content/uploads/2009/05/final-paper_donner_et_al.pdf.

83. Katy E. Pearce, Janine S. Slaker, and Nida Ahmad, "Transnational Families in Armenia and Information Communication Technology," *International Journal of Communication* 7 (2013): 2128–2156.

84. Jan Koum, "WhatsApp," Facebook post, January 6, 2015, https://www.facebook .com/jan.koum/posts/10152994719980011?pnref=story.

85. GlobalWebIndex, "WeChat Dominates Mobile Messaging in APAC," January 2015, http://insight.globalwebindex.net/chart-of-the-day-wechat-dominates-mobile -messaging-in-apac; Meeker, "Internet Trends 2014," 133.

86. Jan van Dijk, *The Deepening Divide: Inequality in the Information Society* (Newbury Park, CA: Sage, 2005); Pippa Norris, *Digital Divide: Civic Engagement, Information Poverty, and the Internet Worldwide* (New York: Cambridge University Press, 2001); NTIA, *Falling through the Net: A Survey of the "Have-Nots" in Rural and Urban America* (Washington, DC: United States Department of Commerce, 1995).

87. Ronald E. Rice and James E. Katz, "Comparing Internet and Mobile Phone Usage: Digital Divides of Usage, Adoption, and Dropouts," *Telecommunications Policy* 27, no. 8–9 (September 2003): 597–623, doi:10.1016/S0308-5961(03)00068-5; Alexander van Deursen and Jan van Dijk, "The Digital Divide Shifts to Differences in Usage," *New Media & Society* 16, no. 3 (June 7, 2013): 507–526, doi:10.1177/1461444813487959; Sinikka Sassi, "Cultural Differentiation or Social Segregation? Four Approaches to the Digital Divide," *New Media & Society* 7, no. 5 (October 01, 2005): 684–700, doi:10.1177/1461444805056012; James, *Digital Interactions in Developing Countries: An Economic Perspective* (New York: Routledge, 2013).

88. Jan van Dijk and Kenneth L. Hacker, "The Digital Divide as a Complex and Dynamic Phenomenon," *The Information Society* 19, no. 4 (September 1, 2003): 315–326, doi:10.1080/01972240309487; Amanda Lenhart and John B. Horrigan, "Re-Visualizing the Digital Divide as a Digital Spectrum," *IT & Society* 1, no. 5 (2003): 23–39; Eszter Hargittai and Yuli Patrick Hsieh, "Digital Inequality," in *The Oxford Handbook of Internet Studies*, ed. William H. Dutton (Oxford, UK, 2013), 129–150; Mark Warschauer, *Technology and Social Inclusion: Rethinking the Digital Divide* (Cambridge, MA: MIT Press, 2003).

89. Raul L. Katz and Hernan Galperin, "Addressing the Broadband Demand Gap: Drivers and Public Policies," SSRN (December 28, 2012).

90. Susan P. Wyche and Laura L. Murphy, "'Dead China-Make' Phones off the Grid: Investigating and Designing for Mobile Phone Use in Rural Africa," in *Proceedings of*

the Designing Interactive Systems Conference on—DIS '12 (New York: ACM Press, 2012), 186–195, doi:10.1145/2317956.2317985.

Chapter 4

1. Everett M. Rogers, *Diffusion of Innovations*, 5th ed. (New York: Free Press, 2003), 219–266.

2. Ibid., 263.

3. James J. Gibson, *The Ecological Approach to Visual Perception* (Boston: Houghton Mifflin, 1979).

4. Donald A. Norman, *The Psychology of Everyday Things* (New York: Basic Books, 1988).

5. Ibid., 9.

6. Norman, *The Psychology of Everyday Things*.

7. Victor Kaptelinin and Bonnie A. Nardi, "Affordances in HCI: Toward a Mediated Action Perspective," in *Proceedings of the 2012 ACM Annual Conference on Human Factors in Computing Systems—CHI '12* (New York: ACM Press, 2012), 967–976, doi:10.1145/2207676.2208541.

8. Michael Chan, "Mobile Phones and the Good Life: Examining the Relationships among Mobile Use, Social Capital and Subjective Well-Being," *New Media & Society* 17, no. 1 (2015): 96–113, doi:10.1177/1461444813516836; Arnau Monterde and John Postill, "Mobile Ensembles: The Uses of Mobile Phones for Social Protest by Spain's Indignados," in *Routledge Companion to Mobile Media*, ed. Gerard Goggin and Larissa Hjorth (New York: Routledge, 2014), 429–438.

9. Barry Wellman, Anabel Quan-haase, Jeffrey Boase, Wenhong Chen, Keith Hampton, Isabel Díaz, and Kakuko Miyata, "The Social Affordances of the Internet for Networked Individualism," *Journal of Computer-Mediated Communication* 8, no. 3 (June 23, 2006), doi:10.1111/j.1083-6101.2003.tb00216.x.

10. Jeffrey W. Treem and Paul M. Leonardi, "Social Media Use in Organizations: Exploring the Affordances of Visibility, Editability, Persistence, and Association," *Communication Yearbook* 36 (2012): 143–189.

11. Daniel Miller, "Materiality: An Introduction," in *Materiality*, ed. Daniel Miller (Chapel Hill, NC: Duke University Press, 2005), 1–50.

12. Jenna Burrell, *Invisible Users: Youth in the Internet Cafés of Urban Ghana* (Cambridge, MA: MIT Press, 2012), 11.

13. Ibid.

14. N. Katherine Hayles, "Print Is Flat, Code Is Deep: The Importance of Media-Specific Analysis," *Poetics Today* 25, no. 1 (March 1, 2004): 67–90, doi:10.1215/03335372-25-1-67; M. McLuhan, *Understanding Media* (New York: McGraw-Hill, 1964); Brian Larkin, *Signal and Noise: Media, Infrastructure, and Urban Culture in Nigeria* (Chapel Hill, NC: Duke University Press, 2008).

15. Becky P. Y. Loo and Y. L. Ngan, "Developing Mobile Telecommunications to Narrow Digital Divide in Developing Countries? Some Lessons from China," *Telecommunications Policy* 36 (2012): 888–900, doi:10.1016/j.telpol.2012.07.015.

16. Jonathan Donner, Shikoh Gitau, and Gary Marsden, "Exploring Mobile-Only Internet Use: Results of a Training Study in Urban South Africa," *International Journal of Communication* 5 (2011): 574–597.

17. Kas Kalba, "The Adoption of Mobile Phones in Emerging Markets: Global Diffusion and the Rural Challenge," *International Journal of Communication* 2 (2008): 631–661.

18. Roxana Barrantes and Hernán Galperin, "Can the Poor Afford Mobile Telephony? Evidence from Latin America," *Telecommunications Policy* 32, no. 8 (2008): 521–530, doi:10.1016/j.telpol.2008.06.002; Ayesha Zainudeen, Nirmali Sivapragasam, Harsha De Silva, Tahani Iqbal, and Dimuthu Ratnadiwakara, "Teleuse at the Bottom of the Pyramid: Findings from a Five Country Study," *SSRN eLibrary* (SSRN, 2005), doi:10.2139/ssrn.1558865.

19. Marshini Chetty, Richard Banks, A. J. Bernheim Brush, Jonathan Donner, and Rebecca E. Grinter, "'You're Capped!' Understanding the Effects of Bandwidth Caps on Broadband Use in the Home," in *Proceedings of CHI 2012, May 5–10, 2010, Austin, TX* (New York: ACM, 2012), 3021–3030, doi:10.1145/2207676.2208714.

20. Nielsen, "Pay-As-You-Phone: How Global Consumers Pay for Mobile," *Nielsen Newswire*, February 19, 2013, http://www.nielsen.com/us/en/insights/news/2013/how-global-consumers-pay-for-mobile.html.

21. GSMA Mobile for Development Intelligence, *Scaling Mobile for Development: Harness the Opportunity* (London, August 2013), 45, https://gsmaintelligence.com/research/?file=130828-scaling-mobile.pdf.

22. Francisco J. Proenza, "The Road to Broadband Development in Developing Countries Is through Competition Driven by Wireless and Internet Telephony," *Information Technologies and International Development* 3, no. 2 (2007): 21–39.

23. Richard Ling, *Taken for Grantedness* (Cambridge, MA: MIT Press, 2012); Kenneth J. Gergen, "The Challenge of Absent Presence," in *Perpetual Contact: Mobile Communication, Private Talk, Public Performance*, ed. James E. Katz and Mark A. Aakhus (Cambridge: Cambridge University Press, 2002), 227–241; Barry Wellman, "Physical Place and Cyberplace: The Rise of Personalized Networking," *International Journal of Urban and Regional Research* 25, no. 2 (June 2001): 227–252, doi:10.1111/1468-2427.00309.

24. Raul Pertierra, *Transforming Technologies: Altered States—Mobile Phone and Internet Use in the Philippines* (Manila: De La Salle University Press, 2006); Mizuko Ito, Daisuke Okabe, and Misa Matsuda, "Personal, Portable, Pedestrian: Mobile Phones in Japanese Life" (Cambridge, MA: MIT Press, 2005); Diana Gant and Sara Kiesler, "Blurring the Boundaries: Cell Phones, Mobility, and the Line between Work and Personal Life," in *Wireless World: Social and Interactional Aspects of the Mobile Age*, ed. Barry Brown, Nicola Green, and Richard Harper (London: Springer, 2001), 121–132, doi:10.1007/978-1-4471-0665-4_9; James E. Katz and Mark Aakhus, "Conclusion: Making Meaning of Mobiles: A Theory of Apparatgeist," in *Perpetual Contact: Mobile Communication, Private Talk, and Public Performance*, ed. James E. Katz and Mark Aakhus (Cambridge, UK: Cambridge University Press, 2002).

25. Richard Ling, *The Mobile Connection: The Cell Phone's Impact on Society* (San Francisco: Morgan Kaufmann, 2004); Manuel Castells, Mireia Fernández-Ardèvol, Jack Linchuan Qiu, and Araba Sey, *Mobile Communication and Society: A Global Perspective* (Cambridge, MA: MIT Press, 2007).

26. Rogers, *Diffusion of Innovations*, 263.

27. Jacqueline Hamilton, "Are Main Lines and Mobile Phones Substitutes or Complements? Evidence from Africa" 27 (2003): 109–133, doi:10.1016/S0308-5961(02)00089-7.

28. Katz and Aakhus, "Conclusion: Making Meaning of Mobiles."

29. Castells et al., *Mobile Communication and Society*, 77.

30. Richard Ling and Brigitte Yttri, "Hyper-Coordination via Mobile Phones in Norway," in Katz and Aakhus, *Perpetual Contact*, 139–169.

31. Ethan Zuckerman, "The Connection between Cute Cats and Web Censorship," . . . *My Heart's in Accra*, June 16, 2007, http://www.ethanzuckerman.com/blog/2007/07/16/the-connection-between-cute-cats-and-web-censorship/.

32. Jeff Grubb, "Minecraft and Candy Crush Saga Dominate Apple's iOS Charts for 2013," *VentureBeat*, December 17, 2013, http://venturebeat.com/2013/12/17/minecraft-and-candy-crush-saga-dominate-apples-ios-charts-for-2013/.

33. Simon Khalaf, "Flurry Five-Year Report: It's an App World. The Web Just Lives in It," *Flurry Blog*, April 3, 2013, http://flurrymobile.tumblr.com/post/115188952445/flurry-five-year-report-its-an-app-world-the.

34. Ingrid Lunden, "Flurry: China Accounts For 24% Of The World's Connected Devices, With 261.3M Active Smartphones And Tablets," *TechCrunch*, July 23, 2013, http://techcrunch.com/2013/07/23/flurry-china-accounts-for-24-of-the-worlds-connected-devices-with-261-3m-active-smartphones-and-tablets/.

35. David Souter, "Mobile Internet Usage and Demand in Kenya: The Experience of Early Adopters," *Making Broadband Accessible for All* (London: Vodafone Group,

May 2011), http://www.vodafone.com/content/dam/vodafone/about/public_policy/ policy_papers/public_policy_series_12.pdf.

36. Ben Goldsmith, "The Smartphone App Economy and App Ecosystems," in Goggin and Hjorth, *Routledge Companion to Mobile Media*, 171–180.

37. Khalaf, "Flurry Five-Year Report."

38. Bryan Pon, Timo Seppälä, and Martin Kenney, "Android and the Demise of Operating System–Based Power: Firm Strategy and Platform Control in the Post-PC World," *Telecommunications Policy* 38, no. 11 (2014): 979–991, doi:10.1016/ j.telpol.2014.05.001.

39. Chris Anderson and Michael Wolff, "The Web Is Dead. Long Live the Internet," *Wired*, September 2010, http://www.wired.com/magazine/2010/08/ff_webrip/; Jonathan Zittrain, *The Future of the Internet and How to Stop It* (New Haven, CT: Yale University Press, 2008).

40. Loo and Ngan, "Developing Mobile Telecommunications to Narrow Digital Divide in Developing Countries?"

41. Paul D. Miller and Svitlana Matviyenko, "Introduction," in *The Imaginary App* (Cambridge, MA: MIT Press, 2014), xi.

42. Gitau, Marsden, and Donner, "After Access"; Arthur Goldstuck, "The Mobile Internet Pinned Down," *World Wide Worx Reports* (Johannesburg, South Africa, May 27, 2010), http://www.worldwideworx.com/the-mobile-internet-pinned-down/.

43. Johnny Ryan, *A History of the Internet and the Digital Future* (London: Reaktion Books, 2010).

44. Lucia Terrenghi, Laura Garcia-Barrio, and Lidia Oshlyansky, "Tablets Use in Emerging Markets: An Exploration," in *Proceedings of the 15th International Conference on Human–Computer Interaction with Mobile Devices and Services—MobileHCI '13* (New York: ACM Press, 2013), 594–599, doi:10.1145/2493190.2494438.

45. John Brownlee, "On Sale for $150: One Laptop Per Child Is Now a Touch Screen Tablet," *Fast Company Design*, July 16, 2013, http://www.fastcodesign.com/1673029/ on-sale-for-150-one-laptop-per-child-is-now-a-touchscreen-tablet-1; Preeti Mudliar and Joyojeet Pal, "ICTD in the Popular Press," in *Proceedings of the Sixth International Conference on Information and Communication Technologies and Development Full Papers—ICTD '13*, vol. 1 (New York: ACM Press, 2013), 43–54, doi:10.1145/2516604 .2516629.

46. Ingrid Lunden, "2014 Tablet Growth Revised Down to 12% on Longer Ownership Cycle, Phablet Popularity," *Techcrunch*, May 29, 2014, http://techcrunch .com/2014/05/29/2014v-tablet-growth-revised-down-to-12-on-longer-ownership -cycle-phablet-rise/.

47. Ericsson, "Ericsson Mobility Report: On the Pulse of the Networked Society" (Stockholm, November 2014), http://www.ericsson.com/res/docs/2014/ericsson-mobility-report-november-2014.pdf.

48. Torsten J. Gerpott, Sandra Thomas, and Michael Weichert, "Personal Characteristics and Mobile Internet Use Intensity of Consumers with Computer-Centric Communication Devices: An Exploratory Empirical Study of iPad and Laptop Users in Germany," *Telematics and Informatics* 30, no. 2 (May 2013): 87–99, doi:10.1016/j.tele.2012.03.008.

49. Katy E. Pearce, Janine Slaker, and Nida Ahmad use a similar term in their conference presentation, "Is Your Web Everyone's Web? Theorizing the Web through the Lens of the Device Divide," Theorizing the Web 2012, April 12, 2012, http://fr.slideshare.net/katyp1/katy-pearce-ttw12. See also Lynn Schofield Clark, Christof Demont-Heinrich, and Scott Webber, "Parents, ICTs, and Children's Prospects for Success: Interviews along the Digital 'Access Rainbow,'" *Critical Studies in Media Communication* 22, no. 5 (2005): 409–426, doi:10.1080/07393180500342985.

Chapter 5

1. Andrew Herman, Jan Hadlaw, and Thom Swiss, "Introduction: Theories of the Mobile Internet: Materialities and Imaginaries," in *Theories of the Mobile Internet: Materialities and Imaginaries*, ed. Andrew Herman, Jan Hadlaw, and Thom Swiss (New York: Routledge, 2015), 1–11.

2. Andrew Feenberg, "Subversive Rationalization: Technology, Power, and Democracy 1," *Inquiry* 35, no. 3–4 (September 1992): 301–322, doi:10.1080/00201749208602296; Merritt Roe Smith and Leo Marx, eds., *Does Technology Drive History? The Dilemma of Technological Determinism* (Cambridge, MA: MIT Press, 1994).

3. D. J. Gunkel, "Second Thoughts: Toward a Critique of the Digital Divide," *New Media & Society* 5, no. 4 (December 1, 2003): 499–522, doi:10.1177/146144480354003; Robin Mansell, "The Information Society and ICT Policy: A Critique of the Mainstream Vision and an Alternative Research Framework," *Journal of Information, Communication and Ethics in Society* 8, no. 1 (2010): 22–41, doi:10.1108/14779961011024792.

4. Janaki Srinivasan and Jenna Burrell, "Revisiting the Fishers of Kerala, India," in *Proceedings of the Sixth International Conference on Information and Communication Technologies and Development: Full Papers—ICTD2015, Vol. 1.* (New York: ACM Press, 2013), 56–66, doi:10.1145/2516604.2516618.

5. Hector Postigo, "Questioning the Web 2.0 Discourse: Social Roles, Production, Values, and the Case of the Human Rights Portal," *The Information Society* 27, no. 3

(May 2011): 181–193, doi:10.1080/01972243.2011.566759; Nicola Green, Richard Harper, G. Murtagh, and Geoff Cooper, "Configuring the Mobile User: Sociological and Industry Views," *Personal and Ubiquitous Computing* 5, no. 2 (July 1, 2001): 146–156, doi:10.1007/s007790170017.

6. Genevieve Bell, "The Age of the Thumb: A Cultural Reading of Mobile Technologies from Asia," in *Thumb Culture*, ed. Peter Glotz, Stefan Bertschi, and Chris Locke (Bielefeld, Germany: Transcript Verlag, 2005), 67–88.

7. François Bar, Francis Pisani, and Matthew Weber, "Mobile Technology Appropriation in a Distant Mirror: Baroque Infiltration, Creolization and Cannibalism," paper presented at Seminario Sobre Desarrollo Economico, Desarrollo Social Y Communicatciones Moviles in America Latina, April 2007, http://www.researchgate.net/profile/Francois_Bar/publication/253284230_Mobile_technology_appropriation _in_a_distant_mirror_baroque_infiltration_creolization_and_cannibalism/links/00b4952f9c4a468a23000000.pdf; Araba Sey, "'We Use It Different': Making Sense of Trends in Mobile Phone Use in Ghana," *New Media & Society* 13, no. 3 (2011), 375–390; Wallace Chuma, "The Social Meanings of Mobile Phones among South Africa's 'Digital Natives': A Case Study," *Media, Culture & Society* 36, no. 3 (March 20, 2014): 398–408, doi:10.1177/0163443713517482.

8. Herman Wasserman, "Mobile Phones, Popular Media, and Everyday African Democracy: Transmissions and Transgressions," *Popular Communication* 9, no. 2 (April 29, 2011): 146–158, doi:10.1080/15405702.2011.562097; Innocentia J. Mhlambi, "'Seyiyakhuluma': IsiZulu as a New Language for Political and Corporate Mass Communication through Mobile Telephony," *African Identities* 10, no. 2 (May 2012): 129–142, doi:10.1080/14725843.2012.657830.

9. Cara J. Wallis, "New Media Practices in China: Youth Patterns, Processes, and Politics," *International Journal of Communication* 5 (2011): 406–436.

10. Gerardine DeSanctis and Marshall Scott Poole, "Capturing the Complexity in Advanced Technology Use: Adaptive Structuration Theory," *Organization Science* 5, no. 2 (1994): 121–147; Richard Heeks, "What Did Giddens and Latour Ever Do for Us? Academic Writings on Information Systems and Development," *DIG Short Papers* (Manchester, UK: Development Informatics Group, University of Manchester, 2001).

11. Sarah Chiumbu, "Exploring Mobile Phone Practices in Social Movements in South Africa—the Western Cape Anti-Eviction Campaign," *African Identities* 10, no. 2 (May 2012): 193–206, doi:10.1080/14725843.2012.657863.

12. Robert W. McChesney, *Digital Disconnect: How Capitalism Is Turning the Internet against Democracy* (New York: The New Press, 2013).

13. Manuel Castells, *The Rise of the Network Society* (Malden, MA: Blackwell Publishing, 1996).

14. Kentaro Toyama, "Technology as Amplifier in International Development," in *Proceedings of the 2011 iConference* (New York: ACM Press, 2011), 75–82, doi:10.1145/1940761.1940772; Evgeny Morozov, *To Save Everything, Click Here: The Folly of Technological Solutionism* (New York: Public Affairs, 2013).

15. Barry Wellman, "Physical Place and Cyberplace: The Rise of Personalized Networking," *International Journal of Urban and Regional Research* 25, no. 2 (June 2001): 227–252, doi:10.1111/1468-2427.00309.

16. Castells, *The Rise of the Network Society.*

17. Ithiel de Sola Pool, "Introduction," *The Social Impact of the Telephone,* ed. Ithiel de Sola Pool (Cambridge, MA: MIT Press, 1977), 1–9; Claude S. Fischer, *America Calling: A Social History of the Telephone to 1940* (Berkeley: University of California Press, 1992).

18. James R. Beniger, *The Control Revolution: Technological and Economic Origins of the Information Society* (Cambridge, MA: Harvard University Press, 1986).

19. Robert J. Saunders, Jeremy J. Warford, and Björn Wellenieus, *Telecommunications and Economic Development,* 2nd ed. (Baltimore, MD: Johns Hopkins University Press, 1994).

20. Dan Gillmor, *We the Media: Grassroots Journalism by the People, for the People* (Sebastopol, CA: O'Reilly, 2004), ch. 2.

21. Richard A. Duncombe and Richard Heeks, "Information and Communication Technologies and Small Enterprise in Africa: Findings from Botswana," in *Small-Scale Enterprises in Developing and Transitional Economies,* ed. Homi Katrak and Roger Strange (Basingstoke, UK: Palgrave, 1999), 285–304.

22. Martin Hilbert and Priscila López, "The World's Technological Capacity to Store, Communicate, and Compute Information," *Science* 332, no. 6025 (April 1, 2011): 60–65, doi:10.1126/science.1200970. Paul Craven and Barry Wellman, "The Network City," *Sociological Inquiry* 43, no. 3–4 (July 1973): 57–88, doi:10.1111/j.1475-682X.1973.tb00003.x; Lee Rainie and Barry Wellman, *Networked: The New Social Operating System* (Cambridge, MA: MIT Press, 2012).

23. Andrew Blum, *Tubes: A Journey to the Center of the Internet* (New York: Ecco, 2012).

24. Daniel Bell, *The Coming of the Post-Industrial Society: A Venture in Social Forecasting* (New York: Basic Books, 1976).

25. Wilson P. Dizard, "The Coming Information Age," *The Information Society* 1, no. 2 (January 1981): 91–112, doi:10.1080/01972243.1981.9959943.

26. Manuel Castells, *The Internet Galaxy: Reflections on the Internet, Business, and Society* (Oxford; New York: Oxford University Press, 2001); Castells, *The Rise of the*

Network Society. Note that in *Communication Power*, Castells describes switches as "sites of concentrated power." Manuel Castells, *Communication Power* (New York: Oxford University Press, 2009), 47. I am using "switch" in a different way.

27. Wellman, "Physical Place and Cyberplace."

28. Yochai Benkler, *The Wealth of Networks: How Social Production Transforms Markets and Freedom* (New Haven, CT: Yale University Press, 2006).

29. At the risk of undermining the explanatory value of the table, it is also worth noting that while the Internet is by far the most important platform for flexible information exchange mediated by digital systems, it is, conceptually, not the only one. From voice paging and voicemail to France's Minitel and Japan's i-mode, many other systems offered read-write functionality that we have come to equate with the Internet; some still do. See, for example, S. Rafaeli, "The Electronic Bulletin Board: A Computer-Driven Mass Medium," *Social Science Computer Review* 2, no. 3 (1984): 123–136, doi:10.1177/089443938600200302; J. Carey and M. C. J. Elton, "The Other Path to the Web: The Forgotten Role of Videotex and Other Early Online Services," *New Media & Society* 11, no. 1–2 (February 1, 2009): 241–260, doi:10.1177/1461444808099576.; and Gerard Goggin and Christina Spurgeon, "Premium Rate Culture: The New Business of Mobile Interactivity," *New Media & Society* 9, no. 5 (October 1, 2007): 753–770, doi:10.1177/1461444807080340.

30. Xu Yan, "The Impact of the Regulatory Framework on Fixed-Mobile Interconnection Settlements: The Case of China and Hong Kong," *Telecommunications Policy* 25, no. 7 (August 2001): 515–532, doi:10.1016/S0308-5961(01)00023-4.

31. Ran Wei and Ven-Hwei Lo, "Staying Connected While on the Move: Cell Phone Use and Social Connectedness," *New Media & Society* 8, no. 1 (2006): 53–72; Wellman, "Physical Place and Cyberplace."

32. James E. Katz and Mark Aakhus, eds., *Perpetual Contact: Mobile Communication, Private Talk, Public Performance* (Cambridge, UK: Cambridge University Press, 2002).

33. Howard Rheingold, *Smart Mobs: The Next Social Revolution* (Cambridge, MA: Perseus Books, 2002).

34. Mizuko Ito, Daisuke Okabe, and Misa Matsuda, *Personal, Portable, Pedestrian: Mobile Phones in Japanese Life* (Cambridge, MA: MIT Press, 2005).

35. Rich Ling and Scott W. Campbell, eds., *The Reconstruction of Space and Time: Mobile Communication Practices* (New Brunswick, NJ: Transaction Publishers, 2009).

36. Henry M. Boettinger, "Our Sixth-and-a-Half Sense," in *The Social Impact of the Telephone*, ed. Ithiel de Sola Pool (Cambridge, MA: MIT Press, 1977), 200–207; Phoebe Sengers, "What I Learned on Change Islands," *Interactions* 18, no. 2 (March 1, 2011): 40–48, doi:10.1145/1925820.1925830.

37. Nicola Green, "On the Move: Technology, Mobility, and the Mediation of Social Time and Space," *The Information Society* 18, no. 4 (July 2002): 281–292, doi:10.1080/01972240290075129.

38. Adriana de Souza e Silva and Jordan Frith, *Mobile Interfaces in Public Spaces: Locational Privacy, Control, and Urban Sociability* (Abingdon, UK: Routledge, 2012).

39. James E. Katz and Mark Aakhus, "Conclusion: Making Meaning of Mobiles: A Theory of Apparatgeist," in *Perpetual Contact: Mobile Communication, Private Talk, and Public Performance*, ed. James E. Katz and Mark Aakhus (Cambridge, UK: Cambridge University Press, 2002).

40. Jason Farman, *Mobile Interface Theory* (Oxford: Routledge, 2012); Mimi Sheller, "Mobile Publics: Beyond the Network Perspective," *Environment and Planning D: Society and Space* 22, no. 1 (2004): 39–52, doi:10.1068/d324t.

41. Diana Gant and Sara Kiesler, "Blurring the Boundaries: Cell Phones, Mobility, and the Line between Work and Personal Life," in *Wireless World: Social and Interactional Aspects of the Mobile Age*, ed. Barry Brown, Nicola Green, and Richard Harper (London: Springer, 2001), 121–132, doi:10.1007/978-1-4471-0665-4_9; Noelle Chesley, "Blurring Boundaries? Linking Technology Use, Spillover, Individual Distress, and Family Satisfaction," *Journal of Marriage and Family* 67, no. 5 (December 2005): 1237–1248, doi:10.1111/j.1741-3737.2005.00213.x.

42. Kenneth J. Gergen, "The Challenge of Absent Presence," in *Perpetual Contact: Mobile Communication, Private Talk, Public Performance*, ed. James E. Katz and Mark A. Aakhus (Cambridge: Cambridge University Press, 2002), 227–241; Christian Licoppe, "'Connected' Presence: The Emergence of a New Repertoire for Managing Social Relationships in a Changing Communication Technoscape," *Environment and Planning D: Society and Space* 22, no. 1 (2004): 135–156, doi:10.1068/d323t.

43. Richard Ling, *Taken for Grantedness* (Cambridge, MA: MIT Press, 2012).

44. Manuel Castells, Mireia Fernández-Ardèvol, Jack Linchuan Qiu, and Araba Sey, *Mobile Communication and Society: A Global Perspective* (Cambridge, MA: MIT Press, 2007), 171–172. Quote references Leopoldina Fortunati, "Mobile Telephone and the Presentation of Self," in *Mobile Communications: Re-Negotiation of the Social Sphere*, ed. Rich Ling and Per E. Pedersen (London: Springer, 2005), 203–218, doi:10.1007/1-84628-248-9_13.

45. Farman, *Mobile Interface Theory*.

46. Mimi Sheller, "Mobile Art: Out of Your Pocket," in Goggin and Hjorth, *Routledge Companion to Mobile Media*, 197–205.

47. Adriana de Souza e Silva, "From Cyber to Hybrid: Mobile Technologies as Interfaces of Hybrid Spaces," *Space and Culture* 9, no. 3 (August 1, 2006): 261–278, doi:10.1177/1206331206289022.

48. Kayzs Varnelis and Anne Friedberg, "Place: Networked Place," in *Networked Publics*, ed. Kazys Varnelis (Cambridge, MA: MIT Press, 2006), 15–42.

49. Jason Farman, "Storytelling with Mobile Media," in Goggin and Hjorth, *Routledge Companion to Mobile Media*, 528.

50. Fischer, *America Calling*.

51. Jonathan Donner, "Blurring Livelihoods and Lives: The Social Uses of Mobile Phones and Socioeconomic Development," *Innovations: Technology, Governance, Globalization* 4, no. 1 (2009): 91–101.

52. Ethan Zuckerman, "The Connection between Cute Cats and Web Censorship," . . . *My Heart's in Accra*, June 16, 2007, http://www.ethanzuckerman.com/blog/2007/07/16/the-connection-between-cute-cats-and-web-censorship/.

53. Beth E. Kolko, Emma J. Rose, and Erica J. Johnson, "Communication as Information-Seeking: The Case for Mobile Social Software for Developing Regions," in *WWW '07 Proceedings of the 16th International Conference on World Wide Web* (New York: ACM Press, 2007), 863–872, doi:10.1145/1242572.1242689; Payal Arora, "The Leisure Divide: Can the 'Third World' Come out to Play?," *Information Development* 28, no. 2 (February 2, 2012): 93–101, doi:10.1177/0266666911433607; Allen E. Brown and Gerald G. Grant, "Highlighting the Duality of the ICT and Development Research Agenda," *Information Technology for Development* 16, no. 2 (April 1, 2010): 96–111, doi:10.1080/02681101003687793; Jonathan Donner, "Research Approaches to Mobile Use in the Developing World: A Review of the Literature," *The Information Society* 24, no. 3 (2008): 140–159, doi:10.1080/01972240802019970; Neha Kumar, "Facebook for Self-Empowerment? A Study of Facebook Adoption in Urban India," *New Media & Society*, July 25, 2014, doi:10.1177/1461444814543999.

54. Donner, "Blurring Livelihoods and Lives."

55. Beth E. Kolko and Cynthia Putnam, "Computer Games in the Developing World: The Value of Non-Instrumental Engagement with ICTs, or Taking Play Seriously," *2009 International Conference on Information and Communication Technologies and Development ICTD*, April 2009, 46–55, doi:10.1109/ICTD.2009.5426705.

56. Jo Tacchi, "Being Meaningfully Mobile: Mobile Phones and Development," in *Technological Determinism and Social Change: Communication in a Tech-Mad World*, ed. Jan Servaes (Lanham, MD: Lexington Books, 2014), 105–124. See also Don Slater, *New Media, Development & Globalization* (Cambridge, UK: Polity Press, 2013).

57. Marlon Parker, Julia Wills, Lucille Aanhuizen, Lester Gilbert, and Gary Wills, "Mobile Instant Messaging Used to Provide Support and Advice to South African Youth," *International Journal of ICT Research and Development in Africa* 3, no. 2 (June 2012): 13–31, doi:10.4018/jictrda.2012070102.

58. Richard Heeks, "Mobiles for Impoverishment?," *ICTs for Development Blog*, December 27, 2008, http://ict4dblog.wordpress.com/2008/12/27/mobiles-for -impoverishment/.

59. Erik Hersman, "Confusing ICT4D Practice with the Tech That Is Used," *Ushahidi Blog* (Nairobi, April 7, 2010), http://blog.ushahidi.com/index.php/2010/04/07/ confusing-ict4d-practice-with-the-tech-that-is-used/; Ken Banks, "An Inconvenient Truth?," *Kiwanja.net*, December 12, 2012, http://www.kiwanja.net/blog/2012/12/ an-inconvenient-truth/.

60. Ory Okolloh, "Ushahidi, or 'Testimony': Web 2.0 Tools for Crowdsourcing Crisis Information," *Participatory Learning and Action* 59, no. 1 (2009): 65–70.

61. Nathan Morrow, Nancy Mock, Adam Papendieck, and Nicholas Kocmich, "Independent Evaluation of the Ushahidi Haiti Project," April 12, 2011, https:// sites.google.com/site/haitiushahidieval/news/finalreportindependentevaluation oftheushahidihaitiproject.

62. Vaughn Hester, Aaron Shaw, and Lukas Biewald, "Scalable Crisis Relief: Crowd-sourced SMS Translation and Categorization with Mission 4636," in *Proceedings of the First ACM Symposium on Computing for Development—ACM DEV '10* (New York: ACM Press, 2010), np, doi:10.1145/1926180.1926199.

63. Linnaea Schuttner, Ntazana Sindano, Mathew Theis, Cory Zue, Jessica Joseph, Roma Chilengi, Benjamin H. Chi, Jeffrey S. A. Stringer, and Namwinga Chint, "A Mobile Phone-Based, Community Health Worker Program for Referral, Follow-Up, and Service Outreach in Rural Zambia: Outcomes and Overview," *Telemedicine and E-Health* 20, no. 8 (2014): 721–728, doi:10.1089/tmj.2013.0240.

64. Carl Hartung, Adam Lerer, Yaw Anokwa, Clint Tseng, Waylon Brunette, and Gaetano Borriello, "Open Data Kit," in *Proceedings of the 4th ACM/IEEE International Conference on Information and Communication Technologies and Development—ICTD '10* (New York: ACM Press, 2010), np, doi:10.1145/2369220.2369236.

65. Ole Hanseth, "From Systems and Tools to Networks and Infrastructures—from Design to Cultivation. Towards a Theory of ICT Solutions and Its Design Methodology Implications," Unpublished manuscript, 2002, http://heim.ifi.uio.no/~oleha/ Publications/ib_ISR_3rd_resubm2.html; Terje Aksel Sanner, Lars Kristian Roland, and Kristin Braa, "From Pilot to Scale: Towards an mHealth Typology for Low-Resource Contexts," *Health Policy and Technology* 1, no. 3 (September 2012): 155–164, doi:10.1016/j.hlpt.2012.07.009, p. 3.

66. Amy Cravens, "How New Devices, Networks, and Consumer Habits Will Change the Web Experience," *Gigaom Research*, January 21, 2013, http://research.gigaom .com/report/how-new-devices-networks-and-consumer-habits-will-change-the -web-experience.

67. Karen Holtzblatt, Ilpo Koskinen, Janaki Kumar, David Rondeau, and John Zimmerman, "Design Methods for the Future That Is Now," in *Extended Abstracts of the 32nd Annual ACM Conference on Human Factors in Computing Systems—CHI EA '14* (New York: ACM Press, 2014), 1063–1068, doi:10.1145/2559206.2579401; Scott Gilbertson, "How a New HTML Element Will Make the Web Faster," *Ars Technica*, September 2, 2014, http://arstechnica.com/information-technology/2014/09/how-a-new-html-element-will-make-the-web-faster/2/.

68. Fischer, *America Calling*.

69. Jeffrey D. Sachs, "The Digital War on Poverty," *GuardianUK Online*, August 21, 2008, http://www.earth.columbia.edu/sitefiles/file/SachsWriting/2008/TheGuardian_DigitalWarOnPoverty_08_21_08.pdf.

70. Md Mahfuz Ashraf, Noushin Laila Ansari, Bushra Tahseen Malike, and Barnaly Rashid, "Evaluating the Impact of Mobile Phone Based 'Health Help Line' Service in Rural Bangladesh," in *Proceedings of M4D2010 10–11 November, Kampala, Uganda*, ed. Jakob Svensson and Gudrun Wicander (Karlstad, Sweden: Universitetstryckeriet, 2012), 15–29.

71. Switchboard.org, February 11, 2014, http://www.switchboard.org/.

72. Barney Warf, "Geographies of Global Telephony in the Age of the Internet," *Geoforum* 45 (December 2013): 219–229, doi:10.1016/j.geoforum.2012.11.008.

73. Rebecca Greenfield, "Facebook's New Plan Could Make Your Cellphone Bill Disappear," *The Atlantic Wire*, January 3, 2013, http://www.theatlanticwire.com/technology/2013/01/facebook-messenger-app-voip-calling/60567/.

74. Balancing Act Africa, "The New Business Model—Everything Becomes Data but What Does It Mean for Operators in Africa?," *Balancing Act Africa*, February 1, 2013, http://www.balancingact-africa.com/news/en/issue-no-640/top-story/the-new-business-mod/en.

75. David G. Messerschmitt, "The Convergence of Telecommunications and Computing: What Are the Implications Today?," *Proceedings of the IEEE* 84, no. 8 (1996): 1167–1186, doi:10.1109/5.533962; Mohsen Khalil and Charles Kenny, "The Next Decade of ICT Development: Access, Applications, and the Forces of Convergence," *Information Technologies & International Development* 4, no. 3 (2008): 1–6.

76. Ofcomm, "The Communications Market 2012 (July)" (London, July 2012), http://stakeholders.ofcom.org.uk/market-data-research/market-data/communications-market-reports/cmr12/.

77. Sachs, "The Digital War on Poverty."

78. Andrew Grantham and George Tsekouras, "Information Society: Wireless ICTs' Transformative Potential," *Futures* 36, no. 3 (April 2004): 359–377, doi:10.1016/S0016-3287(03)00066-1.

79. Jeffrey James, *Digital Interactions in Developing Countries: An Economic Perspective* (New York: Routledge, 2013).

Chapter 6

1. Jason Farman, "Storytelling with Mobile Media," in *Routledge Companion to Mobile Media,* ed. Gerard Goggin and Larissa Hjorth (New York: Routledge, 2014), 528.

2. Richard Ling and Jonathan Donner, *Mobile Communication* (Cambridge, UK: Polity, 2009); Kenneth J. Gergen, "The Challenge of Absent Presence," in *Perpetual Contact: Mobile Communication, Private Talk, Public Performance*, ed. James E. Katz and Mark A. Aakhus (Cambridge: Cambridge University Press, 2002), 227–241; James E. Katz and Mark Aakhus, "Introduction: Framing the Issues," in *Perpetual Contact: Mobile Communication, Private Talk, and Public Performance*, ed. James E. Katz and Mark Aakhus (Cambridge, UK: Cambridge University Press, 2002), 1–13; Rowan Wilken, "Mobile Media, Place, and Location," in *Routledge Companion to Mobile Media,* ed. Gerard Goggin and Larissa Hjorth (New York: Routledge, 2014), 514–527.

3. Michael Bull, "The World According to Sound: Investigating the World of Walkman Users," *New Media & Society* 3, no. 2 (June 1, 2001): 179–197, doi:10.1177/14614440122226047.

4. David N. Breslauer, Robi N. Maamari, Neil A. Switz, Wilbur A. Lam, and Daniel A. Fletcher, "Mobile Phone Based Clinical Microscopy for Global Health Applications," *PloS One* 4, no. 7 (January 2009): e6320, doi:10.1371/journal.pone.0006320.

5. Ling and Donner, *Mobile Communication.*

6. Mizuko Ito, "Mobiles and the Appropriation of Place," *Vodafone Receiver* 8 (2003), np, http://academic.evergreen.edu/curricular/evs/readings/itoShort.pdf; Wilken, "Mobile Media, Place, and Location."

7. Fernando Paragas, "Being Mobile with the Mobile: Cellular Telephony and Renegotiations of Public Transport as Public Sphere," in *Mobile Communications, Re-negotiation of the Social Sphere*, eds. Richard Ling and Per E. Pedersen (Vienna: Springer, 2005), 113–129, doi:10.1007/1-84628-248-9_8.

8. Christian Licoppe, "The 'Crisis of the Summons': A Transformation in the Pragmatics of 'Notifications,' from Phone Rings to Instant Messaging," *The Information Society* 26, no. 4 (July 2010): 288–302, doi:10.1080/01972243.2010.489859.

9. Gergen, "The Challenge of Absent Presence."

10. Steven Vertovec, "Cheap Calls: The Social Glue of Migrant Transnationalism" 4, no. 2 (2004): 219–224, doi:doi:10.1111/j.1471-0374.2004.00088.x.

11. Jack Linchuan Qiu, "Working-Class ICTs, Migrants, and Empowerment in South China," *Asian Journal of Communication* 18, no. 4 (December 2008): 333–347,

doi:10.1080/01292980802344232; Fernando Paragas, "Migrant Mobiles: Cellular Telephony, Transnational Spaces, and the Filipino Diaspora," in *A Sense of Place: The Global and the Local in Mobile Communication*, ed. Kristóf Nyíri (Vienna: Passagen Verlag, 2005), 241–249; Mirca Madianou and Daniel Miller, "Mobile Phone Parenting: Reconfiguring Relationships between Filipina Migrant Mothers and Their Left-behind Children," *New Media & Society* 13, no. 3 (March 23, 2011): 457–470, doi:10.1177/1461444810393903; Cara J. Wallis, "(Im)mobile Mobility: Marginal Youth and Mobile Phones in Beijing," in *Mobile Communication: Bringing Us Together and Tearing Us Apart*, ed. Richard Ling and Scott W. Campbell (New Brunswick, NJ: Transaction Publishers, 2011), 61–81; Xueming Lang, Elisa Oreglia, and Suzanne Thomas, "Social Practices and Mobile Phone Use of Young Migrant Workers," in *Proceedings of the 12th International Conference on Human Computer Interaction with Mobile Devices and Services—MobileHCI '10* (New York: ACM Press, 2010), 59–62, doi:10.1145/1851600.1851613.

12. Jenna Burrell and Ken Anderson, "'I Have Great Desires to Look beyond My World': Trajectories of Information and Communication Technology Use among Ghanaians Living Abroad," *New Media & Society* 10, no. 2 (April 1, 2008): 203–224, doi:10.1177/1461444807086472; Tingyu Kang, "Homeland Re-Territorialized: Revisiting the Role of Geographical Places in the Formation of Diasporic Identity in the Digital Age," *Information, Communication & Society* 12, no. 3 (April 2009): 326–343, doi:10.1080/13691180802635448; Harry H. Hiller and Tara M. Franz, "New Ties, Old Ties and Lost Ties: The Use of the Internet in Diaspora," *New Media & Society* 6, no. 6 (December 1, 2004): 731–752, doi:10.1177/146144804044327.

13. Nirmali Sivapragasam, Aileen Agüero, and Harsha De Silva, "The Potential of Mobile Remittances for the Bottom of the Pyramid: Findings from Emerging Asia," *Info* 13, no. 3 (2011): 91–109, doi:10.1108/14636691111131475; Olga Morawczynski, "Surviving in the 'Dual System': How M-PESA Is Fostering Urban-to-Rural Remittances in a Kenyan Slum," in *Proceedings of the IFIP WG 9.4—University of Pretoria Joint Workshop* (IFIP: Pretoria, South Africa, 2008), 110–127, http://researchspace.csir.co.za/dspace/bitstream/10204/2501/1/Phahlamohlaka_2008.pdf#page=110.

14. Katy E. Pearce, Janine S. Slaker, and Nida Ahmad, "Transnational Families in Armenia and Information Communication Technology," *International Journal of Communication* 7 (2013): 2128–2156.

15. Loretta Baldassar, Core Vellekoop Baldock, and Raelene Wilding, *Families Caring Across Borders: Migration, Aging and Transnational Caregiving* (London: Palgrave Macmillan, 2007).

16. Lee Komito and Jessica Bates, "Virtually Local: Social Media and Community among Polish Nationals in Dublin," ed. Ian Cornelius, *Aslib Proceedings* 61, no. 3 (May 22, 2009): 232–244, doi:10.1108/00012530910959790.

17. Jo Tacchi, Kathi R. Kitner, and Kate Crawford, "Meaningful Mobility: Gender, Development and Mobile Phones," *Feminist Media Studies* 12, no. 4 (December 2012): 528–537, doi:10.1080/14680777.2012.741869; Assa Doron, "Mobile Persons: Cell Phones, Gender and the Self in North India," *The Asia Pacific Journal of Anthropology* 13, no. 5 (November 2012): 414–433, doi:10.1080/14442213.2012.726253; Ashima Goyal, "Developing Women: Why Technology Can Help," *Information Technology for Development* 17, no. 2 (April 2011): 112–132, doi:10.1080/02681102.2010. 537252; Nancy J. Hafkin and Sophia Huyer, "Introduction," in *Cinderella Or Cyberella?: Empowering Women in the Knowledge Society*, ed. Sophia Huyer and Nancy J. Hafkin (Bloomfield, CT: Kumarian Press, 2006), 1–14.

18. Hazel Gillard, Debra Howcroft, Natalie Mitev, and Helen Richardson, "'Missing Women': Gender, ICTs, and the Shaping of the Global Economy," *Information Technology for Development* 14, no. 4 (October 2008): 262–279, doi:10.1002/itdj.20098; Ayesha Zainudeen, Tahani Iqbal, and Rohan Samarajiva, "Who's Got the Phone? Gender and the Use of the Telephone at the Bottom of the Pyramid," *New Media & Society* 12, no. 4 (February 09, 2010): 549–566, doi:10.1177/1461444809346721.

19. GSMA, *Bridging the Gender Gap: Mobile Access and Usage in Low- and Middle-Income Countries*, London, UK, 2015, http://www.gsma.com/connectedwomen/ wp-content/uploads/2015/02/GSM0001_02252015_GSMAReport_FINAL-WEB -spreads.pdf.

20. Anne Milek, Christoph Stork, and Alison Gillwald, "Engendering Communication: A Perspective on ICT Access and Usage in Africa," *Info* 13, no. 3 (2011): 125–141, doi:10.1108/14636691111131493.

21. GSMA mWomen, "Striving and Surviving : Exploring the Lives of Women at the Base of the Pyramid," January 27, 2012, http://www.gsma.com/ mobilefordevelopment/gsma-mwomen-striving-and-surviving-exploring-the-lives -of-bop-women; Leslie L. Dodson, S. Revi Sterling, and John K. Bennett, "Minding the Gaps," in *Proceedings of the Sixth International Conference on Information and Communication Technologies and Development Full Papers—ICTD '13*, vol. 1 (New York: ACM Press, 2013), 79–88, doi:10.1145/2516604.2516626.

22. Jenna Burrell, "Evaluating Shared Access: Social Equality and the Circulation of Mobile Phones in Rural Uganda," *Journal of Computer-Mediated Communication* 15, no. 2 (January 2010): 230–250, doi:10.1111/j.1083-6101.2010.01518.x.

23. Arul Chib and Vivian Hsueh-Hua Chen, "Midwives with Mobiles: A Dialectical Perspective on Gender Arising from Technology Introduction in Rural Indonesia," *New Media & Society* 13, no. 3 (March 31, 2011): 486–501, doi:10.1177/ 1461444810393902.

24. Laura L. Murphy and A. E. Priebe, "'My Co-Wife Can Borrow My Mobile Phone!': Gendered Geographies of Cell Phone Usage and Significance for Rural Kenyans,"

Gender, Technology and Development 15, no. 1 (August 5, 2011): 1–23, doi:10.1177/097185241101500101.

25. Sun Sun Lim, "Women, 'Double Work,' and Mobile Media," in Goggin and Hjorth, *Routledge Companion to Mobile Media*, 356–364.

26. Jukka Jouhki, "A Phone of One's Own?," *Suomen Antropologi: Journal of the Finnish Anthropological Society* 38, no. 1 (2013): 37–58.

27. Assa Doron and Robin Jeffrey, *The Great Indian Phone Book: How the Cheap Cell Phone Changes Business, Politics, and Daily Life* (Cambridge, MA: Harvard University Press, 2013), 183.

28. Katy E. Pearce, "The Reproduction and Amplification of Gender Inequality Online: The Case of Azerbaijan," in CHANGE Seminar, University of Washington, October 2013.

29. Jo Tacchi, "Being Meaningfully Mobile: Mobile Phones and Development," in *Technological Determinism and Social Change: Communication in a Tech-Mad World*, ed. Jan Servaes (Lanham, MD: Lexington Books, 2014), 114. See also Preeti Mudliar and Nimmi Rangaswamy, "Offline Strangers, Online Friends," in *Proceedings of the 33rd Annual ACM Conference on Human Factors in Computing Systems—CHI '15* (New York: ACM, 2015), 3799–3808, doi:10.1145/2702123.2702533.

30. K. Balasubramanian, P. Thamizoli, Abdurrahman Umar, and Asha Kanwar, "Using Mobile Phones to Promote Lifelong Learning among Rural Women in Southern India," *Distance Education* 31, no. 2 (August 2010): 193–209, doi:10.1080/01587919.2010.502555.

31. Han Ei Chew, Mark Levy, and Vigneswara Ilavarasan, "The Limited Impact of ICTs on Microenterprise Growth : A Study of Businesses Owned by Women in Urban India," *Information Technologies & International Development* 7, no. 4 (2011): 1–16.

32. Gina Porter, Kate Hampshire, Albert Abane, Alister Munthali, Elsbeth Robson, Mac Mashiri, and Augustine Tanle, "Youth, Mobility and Mobile Phones in Africa: Findings from a Three-Country Study," *Information Technology for Development*, February 3, 2012, 1–18, doi:10.1080/02681102.2011.643210.

33. Antonio M. Battro, "One Laptop Per Child: Comments on Jeffrey James's Critique," *Social Science Computer Review* 31, no. 1 (January 18, 2013): 133–135, doi:10.1177/0894439311421752.

34. Elizabeth Beckmann, "Learners on the Move: Mobile Modalities in Development Studies," *Distance Education* 31, no. 2 (August 2010): 159–173, doi:10.1080/01587919.2010.498081; Yeonjeong Park, "A Pedagogical Framework for Mobile Learning : Categorizing Educational Applications of Mobile Technologies into Four Types," *International Review of Research in Open and Distance Learning* 12, no. 2 (2011): 78–102, doi:10.3394/0380-1330(2006)32.

35. Laurie Butgereit, "Math on MXit: Using MXit as a Medium for Mathematics Education," paper presented at Meraka INNOVATE Conference for Educators, CSIR, April 18–20, 2007 (Pretoria, 2007); Matthew Kam, Anuj Kumar, Shirley Jain, Akhil Mathur, and John Canny, "Improving Literacy in Rural India: Cellphone Games in an After-School Program," in *2009 International Conference on Information and Communication Technologies and Development (ICTD)* (IEEE, 2009), 139–149, doi:10.1109/ICTD.2009.5426712.

36. Sagarmay Deb, "How to Use Mobile Technology to Provide Distance Learning in an Efficient Way Using Advanced Multimedia Tools in Developing Countries," in *Multimedia Computer Graphics and Broadcasting*, ed. Tai-hoon Kim et al., vol. 262 (Berlin: Springer Berlin Heidelberg, 2012), 210–216, doi:10.1007/978-3-642-27204-2_26.

37. Glenn Auld, Ilana Snyder, and Michael Henderson, "Using Mobile Phones as Placed Resources for Literacy Learning in a Remote Indigenous Community in Australia," *Language and Education* 26, no. 4 (July 2012): 279–296, doi:10.1080/09500782.2012.691512.

38. Hatem Said Alismail, Aysha Siddique, Malcolm Frederick Dias, Anthony Velazquez, Mary Beatrice Dias, Sarah M. Belousov, Ermine A. Teves, Rotimi Abimbola, Daniel Nuffer, Bradley Hall, and M. Bernardine Dias,"Combining Web Technology and Mobile Phones to Enhance English Literacy in Underserved Communities," in *Proceedings of the First ACM Symposium on Computing for Development* (New York: ACM Press, 2010), np, doi:10.1145/1926180.1926202.

39. Michael Trucano, "Missing Perspectives on MOOCs: Views from Developing Countries," *World Bank Edutech Blog* (Washington, DC, April 19, 2013), http://blogs.worldbank.org/edutech/MOOC-perspectives.

40. X. Gu, F. Gu, and J. M. Laffey, "Designing a Mobile System for Lifelong Learning on the Move," *Journal of Computer Assisted Learning* 27, no. 3 (June 19, 2011): 204–215, doi:10.1111/j.1365-2729.2010.00391.x.

41. Payal Arora, "The Leisure Divide: Can the 'Third World' Come out to Play?," *Information Development* 28, no. 2 (February 2, 2012): 93–101, doi:10.1177/0266666911433607; Beth E. Kolko and Cynthia Putnam, "Computer Games in the Developing World: The Value of Non-Instrumental Engagement with ICTs, or Taking Play Seriously," *2009 International Conference on Information and Communication Technologies and Development ICTD*, April 2009, 46–55, doi:10.1109/ICTD.2009.5426705.

42. Jenny C. Aker, Christopher Ksoll, and Travis J. Lybbert, "Can Mobile Phones Improve Learning? Evidence from a Field Experiment in Niger," *American Economic Journal: Applied Economics* 4, no. 4 (October 2012): 94–120, doi:10.1257/app.4.4.94.

43. International Assessment of Agricultural Science and Technology for Development, *Agriculture at a Crossroads: Global Report* (Washington, DC: Island Press, 2009).

44. Peer Stein, Tony Goland, and Robert Schiff, "Two Trillion and Counting: Assessing the Credit Gap for Micro, Small, and Medium-Size Enterprises in the Developing World" (Washington, DC, October 2010), http://www.mspartners.org/download/Twotrillion.pdf.

45. Jonathan Donner and Marcela X. Escobari, "A Review of Evidence on Mobile Use by Micro and Small Enterprises in Developing Countries," *Journal of International Development* 22, no. 5 (2010): 641–658, doi:10.1002/jid.1717.

46. Jenny C. Aker, "Dial 'A' for Agriculture: A Review of Information and Communication Technologies for Agricultural Extension in Developing Countries," *Agricultural Economics* 42, no. 6 (November 15, 2011): 631–647, doi:10.1111/j.1574-0862.2011.00545.x; Marcel Fafchamps and Bart Minten, "Impact of SMS-Based Agricultural Information on Indian Farmers," *The World Bank Economic Review* 26, no. 3 (February 27, 2012): 383–414, doi:10.1093/wber/lhr056.

47. Robert J. Saunders, Jeremy J. Warford, and Björn Wellenieus, *Telecommunications and Economic Development*, 2nd ed. (Baltimore, MD: Johns Hopkins University Press, 1994); Karen Eggleston, Robert Jensen, and Richard Zeckhauser, "Information and Communication Technologies, Markets, and Economic Development," in *The Global Information Technology Report 2001–2002*, eds. Geoffrey Kirkman, Peter Cornelius, Jeffrey D. Sachs, and Klaus Schwab (New York: Oxford University Press, 2002), 62–75.

48. Robert Jensen, "The Digital Provide," *Quarterly Journal of Economics* 124, no. 4 (November 2009): 879–924, doi:10.1162/qjec.2009.124.4.ix; Richard A. Duncombe, *Understanding Mobile Phone Impact on Livelihoods in Developing Countries: A New Research Framework*, Development Informatics Working Paper Series #48 (Manchester, UK, 2012); Gina Porter, "Mobile Phones, Livelihoods and the Poor in Sub-Saharan Africa: Review and Prospect," *Geography Compass* 6, no. 5 (May 9, 2012): 241–259, doi:10.1111/j.1749-8198.2012.00484.x; Donner and Escobari, "A Review of Evidence on Mobile Use by Micro and Small Enterprises in Developing Countries"; Naomi J. Halewood and Priya Surya, "Mobilizing the Agricultural Value Chain," in *Information and Communications for Development 2012: Maximizing Mobile*, ed. World Bank (Washington, DC: World Bank, 2012), 31–44; Seth W. Norton, "Transaction Costs, Telecommunications, and the Microeconomics of Macroeconomic Growth," *Economic Development and Cultural Change* 41, no. 1 (October 1992): 175–196, doi:10.1086/452002; Saunders, Warford, and Wellenieus, *Telecommunications and Economic Development*.

49. Megumi Muto and Takashi Yamano, "The Impact of Mobile Phone Coverage Expansion on Market Participation: Panel Data Evidence from Uganda," *World*

Development 37, no. 12 (December 2009): 1887–1896, doi:doi:10.1016/j.worlddev
.2009.05.004.

50. James R. Beniger, *The Control Revolution: Technological and Economic Origins of the Information Society* (Cambridge, MA: Harvard University Press, 1986).

51. Fafchamps and Minten, "Impact of SMS-Based Agricultural Information on Indian Farmers."

52. César Antúnez-de-Mayolo, "The Role of Innovation at the Bottom of the Pyramid in Latin America: Eight Case Studies," *Procedia—Social and Behavioral Sciences* 40 (January 2012): 134–140, doi:10.1016/j.sbspro.2012.03.172.

53. Aker, "Dial 'A' for Agriculture"; Jonathan Donner, "Mobile-Based Livelihood Services in Africa: Pilots and Early Deployments," in *Communication Technologies in Latin America and Africa: A Multidisciplinary Perspective*, ed. Mireia Fernández-Ardèvol and Adela Ros (Barcelona: IN3, 2009), 37–58.

54. Terje Aksel Sanner, Lars Kristian Roland, and Kristin Braa, "From Pilot to Scale: Towards an mHealth Typology for Low-Resource Contexts," *Health Policy and Technology* 1, no. 3 (September 2012): 155–164, doi:10.1016/j.hlpt.2012.07.009.

55. Mark Davies, "Fertilizer by Phone: Esoko Enhances African Farmers' Livelihoods through Innovations in Data Access (Innovations Case Narrative: Esoko)," *Innovations: Technology, Governance, Globalization* 7, no. 4 (October 2012): 27–41, doi:10.1162/INOV_a_00150.

56. Ayesha Zainudeen and Rohan Samarajiva, "CellBazaar: Enabling M-Commerce in Bangladesh," *Information Technologies & International Development* 7, no. 3 (2011): 61–76; Kamal Quadir and Naeem Mohaiemen, "CellBazaar: A Market in Your Pocket," *Innovations: Technology, Governance, Globalization* 4, no. 1 (January 2009): 57–69, doi:10.1162/itgg.2009.4.1.57.

57. Zainudeen and Samarajiva, "CellBazaar: Enabling M-Commerce in Bangladesh."

58. Quentin Hardy, "Cloud Computing for the Poorest Countries," *Bits* (New York Times Technology blog), August 29, 2012, http://bits.blogs.nytimes.com/2012/08/29/cloud-computing-for-the-poorest-countries/.

59. iCow, "What Is iCow?," February 25, 2014, http://www.icow.co.ke/.

60. Susan P. Wyche, Andrea Forte, and Sarita Yardi Schoenebeck, "Hustling Online: Understanding Consolidated Facebook Use in an Informal Settlement in Nairobi," in *Proceedings of the SIGCHI Conference on Human Factors in Computing Systems—CHI '13* (New York: ACM Press, 2013), 2823–2832, doi:10.1145/2470654.2481391; Jonathan Donner and Andrew Maunder, "Beyond the Phone Number: Challenges of Representing Informal Microenterprise on the Internet," in *Living Inside Mobile Social Information*, ed. James E. Katz (Dayton, Ohio: Greyden Press, 2014), 159–192.

61. Sutanuka Ghosal, "Farmers Using Facebook to Discuss Prices and Plan Strategy," *The Economic Times*, February 10, 2012, http://articles.economictimes.indiatimes.com/2012-02-10/news/31046360_1_turmeric-farmers-social-media-sangli-district.

62. Donner and Maunder, "Beyond the Phone Number."

63. Jeffrey R. Brown and Austan Goolsbee, "Does the Internet Make Markets More Competitive? Evidence from the Life Insurance Industry," *Journal of Political Economy* 110, no. 3 (2002): 481–507.

64. Jeffrey F. Rayport and John J. Sviokla, "Managing in the Marketspace," *Harvard Business Review* 72, no. 6 (November–December 1994): 141–150.

65. Maja Andjelkovic and Saori Imaizumi, "Mobile Entrepreneurship and Employment," in *Information and Communications for Development 2012: Maximizing Mobile*, ed. World Bank, vol. 2011 (Washington, DC: World Bank, 2012), 75–86.

66. Amber Houssian, Mohammad Kilany, and Jacob Korenblum, "Mobile Phone Job Services: Linking Developing-Country Youth with Employers, via SMS," in *2009 International Conference on Information and Communication Technologies and Development (ICTD)* (New York: IEEE, 2009), 491, doi:10.1109/ICTD.2009.5426734.

67. Joel Ross, Lilly Irani, M. Six Silberman, Andrew Zaldivar, and Bill Tomlinson, "Who Are the Crowdworkers?," in *Proceedings of the 28th International Conference on Human Factors in Computing Systems—Extended Abstracts CHI EA 2010* (New York: ACM Press, 2010), 2863–2872, doi:10.1145/1753846.1753873.

68. Anand Kulkarni, Philipp Gutheim, Prayag Narula, David Rolnitzky, Tapan Parikh, and Bjorn Hartmann, "MobileWorks: Designing for Quality in a Managed Crowdsourcing Architecture," *IEEE Internet Computing* 16, no. 5 (September 2012): 28–35, doi:10.1109/MIC.2012.72.

69. Duncan Graham-Rowe, "How Nathan Eagle's Jana Service Aims to Top Up Africa," *Wired.co.uk*, August 2, 2011, http://www.wired.co.uk/magazine/archive/2011/09/start/airtime-entrepreneur; Nathan Eagle, "Txteagle: Mobile Crowdsourcing," in *IDGD '09 Proceedings of the 3rd International Conference on Internationalization, Design and Global Development: Held as Part of HCI International 2009* (Heidelberg: Springer-Verlag Berlin, 2009), 447–456, doi:10.1007/978-3-642-02767-3_50.

70. Carol Upadhya, "Controlling Offshore Knowledge Workers: Power and Agency in India's Software Outsourcing Industry," *New Technology, Work and Employment* 24, no. 1 (March 2009): 2–18, doi:10.1111/j.1468-005X.2008.00215.x. See also job services for heterogeneous device environments, like Souktel.org (Jordan), LaborNet.in and Babajob.com (India), and Ummeli.com (South Africa).

71. Ahmed Dermish, Christoph Kneiding, Paul Leishman, and Ignacio Mas, "Branchless and Mobile Banking Solutions for the Poor : A Survey of the Literature,"

Innovations: Technology, Governance, Globalization 6, no. 4 (January 23, 2012): 81–98; Jonathan Donner and Camilo Andres Tellez, "Mobile Banking and Economic Development: Linking Adoption, Impact, and Use," *Asian Journal of Communication* 18, no. 4 (December 2008): 318–332, doi:10.1080/01292980802344190; Kevin P. Donovan, "Mobile Money for Financial Inclusion," in *Information and Communications for Development 2012: Maximizing Mobile*, ed. World Bank (Washington, DC: World Bank, 2012), 61–74; Bill Maurer, "Mobile Money: Communication, Consumption and Change in the Payments Space," *Journal of Development Studies* 48, no. 5 (2012): 1–16, doi:10.1080/00220388.2011.621944; Erin B. Taylor and Heather A. Horst, "The Aesthetics of Mobile Money Platforms in Haiti," in Goggin and Hjorth, *Routledge Companion to Mobile Media*, 462–471.

72. Nick Hughes and Susie Lonie, "M-PESA: Mobile Money for the 'Unbanked' Turning Cellphones into 24-Hour Tellers in Kenya," *Innovations: Technology, Governance, Globalization* 2, no. 1–2 (April 2007): 63–81, doi:10.1162/itgg.2007.2.1-2.63.

Chapter 7

1. Rowan Wilken, "Mobile Media, Place, and Location," in *Routledge Companion to Mobile Media*, ed. Gerard Goggin and Larissa Hjorth (New York: Routledge, 2014), 514–527, p. 522.

2. Ibid., 514.

3. Jason Farman, "Storytelling with Mobile Media," in *Routledge Companion to Mobile Media*, ed. Gerard Goggin and Larissa Hjorth (New York: Routledge, 2014), 528–537, p. 528.

4. James E. Katz and Chih-Hui Lai, "Mobile Locative Media," in *Routledge Companion to Mobile Media*, ed. Gerard Goggin and Larissa Hjorth (New York: Routledge, 2014), 53–62.

5. Daniel Palmer, "Mobile Media Photography," in *Routledge Companion to Mobile Media*, ed. Gerard Goggin and Larissa Hjorth (New York: Routledge, 2014), 245–255.

6. Max Schleser, "A Decade of Mobile Moving-Image Practice," in *Routledge Companion to Mobile Media*, ed. Gerard Goggin and Larissa Hjorth (New York: Routledge, 2014), 157–170.

7. Jonathan Raper, Georg Gartner, Hassan Karimi, and Chris Rizos, "A Critical Evaluation of Location Based Services and Their Potential," *Journal of Location Based Services* 1, no. 1 (March 2007): 5–45, doi:10.1080/17489720701584069.

8. C. Neumayer and G. Stald, "The Mobile Phone in Street Protest: Texting, Tweeting, Tracking, and Tracing," *Mobile Media & Communication* 2, no. 2 (April 16, 2014): 117–133, doi:10.1177/2050157913513255.

9. Ingrid Richardson and Larissa Hjorth, "Mobile Games: From Tetris to Four-square," in *Routledge Companion to Mobile Media*, ed. Gerard Goggin and Larissa Hjorth (New York: Routledge, 2014), 256–266.

10. Farman, "Storytelling with Mobile Media."

11. Kenneth J. Gergen, "The Challenge of Absent Presence," in *Perpetual Contact: Mobile Communication, Private Talk, Public Performance*, ed. James E. Katz and Mark A. Aakhus (Cambridge: Cambridge University Press, 2002), 227–241.

12. Farman, "Storytelling with Mobile Media," 536.

13. Mark Graham, "The Virtual Dimension," in *Global City Challenges: Debating a Concept, Improving the Practice*, ed. M. Acuto and W. Steele (London: Palgrave, 2013), 117–139.

14. Katz and Lai, "Mobile Locative Media."

15. Graham, "The Virtual Dimension."

16. Larissa Hjorth and Kay Gu, "The Place of Emplaced Visualities: A Case Study of Smartphone Visuality and Location-Based Social Media in Shanghai, China," *Continuum* 26, no. 5 (2012): 699–713, doi:10.1080/10304312.2012.706459.

17. Marion Walton, "Social Distance, Mobility and Place: Global and Intimate Genres in Geo-Tagged Photographs of Guguletu, South Africa," in *Proceedings of the 8th ACM Conference on Designing Interactive Systems—DIS '10* (New York: ACM Press, 2010), 35–38, doi:10.1145/1858171.1858178.

18. Jason Farman, *Mobile Interface Theory* (Oxford: Routledge, 2012); Mimi Sheller, "Mobile Publics: Beyond the Network Perspective," *Environment and Planning D: Society and Space* 22, no. 1 (2004): 39–52, doi:10.1068/d324t; Lee Humphreys, "Mobile Social Networks and Urban Public Space," *New Media & Society* 12, no. 5 (February 9, 2010): 763–778, doi:10.1177/1461444809349578; Adriana de Souza e Silva and Daniel M. Sutko, eds., *Digital Cityscapes* (New York: Peter Lang, 2009); Kazys Varnelis and Anne Friedberg, "Place: Networked Place," in *Networked Publics*, ed. Kazys Varnelis (Cambridge, MA: MIT Press, 2006), 15–42.

19. Howard Rheingold, "Mobile Media and Political Collective Action," in *Handbook of Mobile Communication Studies* (Cambridge, MA: MIT Press, 2008), 225–229.

20. Manuel Castells, *Communication Power* (New York: Oxford University Press, 2009); Manuel Castells, *Networks of Outrage and Hope: Social Movements in the Internet Age* (Cambridge, UK: Polity, 2012).

21. W. Lance Bennett and Alexandra Segerberg, "The Logic of Connective Action: Digital Media and the Personalization of Contentious Politics," *Information, Communication & Society* 15, no. 5 (June 2012): 739–768, doi:10.1080/1369118X .2012.670661.

22. Rheingold, *Smart Mobs*.

23. Raul Pertierra, Eduardo F. Ugarte, Alicia Pingol, Joel Hernandez, and Nikos Lexis Dacanay, *Txt-Ing Selves: Cell Phones and Philippine Modernity* (Manila: De La Salle University Press, 2002).

24. Janey Gordon, "The Mobile Phone and the Public Sphere: Mobile Phone Usage in Three Critical Situations," *Convergence: The International Journal of Research into New Media Technologies* 13, no. 3 (August 1, 2007): 307–319, doi:10.1177/1354856507079181; Jack Linchuan Qiu, "Communication & Global Power Shifts 'Power To the People!': Mobiles, Migrants, and Social Movements in Asia," *International Journal of Communication* 8 (2014), 376–391.

25. E.g., Stefania Vicari, "Networks of Contention: The Shape of Online Transnationalism in Early Twenty-First Century Social Movement Coalitions," *Social Movement Studies* 13, no. 1 (September 19, 2013): 92–109, doi:10.1080/14742837.2013.83 2621; Diana Guillén, "Mexican Spring? #YoSoy132, the Emergence of an Unexpected Collective Actor in the National Political Arena," *Social Movement Studies* 12, no. 4 (November 2013): 471–476, doi:10.1080/14742837.2013.830563.

26. Clay Shirky, "The Political Power of Social Media," *Foreign Affairs* 90, no. 1 (2011): 28–41; R. Kelly Garrett, "Protest in an Information Society: A Review of Literature on Social Movements and New ICTs," *Information, Communication & Society* 9, no. 2 (2006): 202–224, doi:10.1080/13691180600630773.New; Neumayer and Stald, "The Mobile Phone in Street Protest."

27. Maarit Makinen and Mary Wangu Kuira, "Social Media and Postelection Crisis in Kenya," *The International Journal of Press/Politics* 13, no. 3 (July 1, 2008): 328–335, doi:10.1177/1940161208319409.

28. Noah Arceneaux and A. Schmitz Weiss, "Seems Stupid until You Try It: Press Coverage of Twitter, 2006–9," *New Media & Society* 12, no. 8 (May 18, 2010): 1262–1279, doi:10.1177/1461444809360773.

29. Arnau Monterde and John Postill, "Mobile Ensembles: The Uses of Mobile Phones for Social Protest by Spain's Indignados," in *Routledge Companion to Mobile Media*, ed. Gerard Goggin and Larissa Hjorth (New York: Routledge, 2014), 429–438.

30. Tahir Abbas, "Political Culture and National Identity in Conceptualising the Gezi Park Movement," *Insight Turkey* 15, no. 4 (2013): 19–27.

31. Irina Khmelko and Yevgen Pereguda, "An Anatomy of Mass Protests: The Orange Revolution and Euromaydan Compared," *Communist and Post-Communist Studies* 47, no. 2 (June 2014): 227–236, doi:10.1016/j.postcomstud.2014.04.013.

32. Nathan Jurgenson, "When Atoms Meet Bits: Social Media, the Mobile Web and Augmented Revolution," *Future Internet* 4, no. 1 (January 23, 2012): 83–91, doi:10.3390/fi4010083.

33. Zeynep Tufekci and Christopher Wilson, "Social Media and the Decision to Participate in Political Protest: Observations from Tahrir Square," *Journal of Communication* 62, no. 2 (April 6, 2012): 363–379, doi:10.1111/j.1460-2466.2012.01629.x; Anita Breuer and Jacob Groshek, "Online Media and Offline Empowerment in Post-Rebellion Tunisia: An Analysis of Internet Use During Democratic Transition," *Journal of Information Technology & Politics* 11, no. 1 (January 2, 2014): 25–44, doi:10.1080/19331681.2013.850464.

34. Monterde and Postill, "Mobile Ensembles," 436.

35. Neumayer and Stald, "The Mobile Phone in Street Protest."

36. Michael Chanan, "Video, Activism and the Art of Small Media," *Transnational Cinemas* 2, no. 2 (March 8, 2012): 217–226, doi:10.1386/trac.2.2.217_7.

37. Jennifer Earl, Heather McKee Hurwitz, Analicia Mejia Mesinas, Margaret Tolan, and Ashley Arlotti, "This Protest Will Be Tweeted: Twitter and Protest Policing during the Pittsburgh G20," *Information, Communication & Society* 16, no. 4 (May 2013): 459–478, doi:10.1080/1369118X.2013.777756; Aaron S. Veenstra, Narayanan Iyer, Mohammad Delwar Hossain, and Jiwoo Park, "Time, Place, Technology: Twitter as an Information Source in the Wisconsin Labor Protests," *Computers in Human Behavior* 31 (February 2014): 65–72, doi:10.1016/j.chb.2013.10.011.

38. K. Hazel Kwon, Yoonjae Nam, and Derek Lackaff, "Wireless Protesters Move Around: Informational and Coordinative Use of Information and Communication Technologies for Protest Politics," *Journal of Information Technology & Politics* 8, no. 4 (October 2011): 383–398, doi:10.1080/19331681.2011.559743.

39. Stuart Allan, "Citizen Journalism and the Rise of 'Mass Self-Communication': Reporting the London Bombings," *Global Media Journal* 1, no. 1 (2007): 1–20; John Mills, Paul Egglestone, Omer Rashid, and Heli Väätäjä, "MoJo in Action: The Use of Mobiles in Conflict, Community, and Cross-Platform Journalism," *Continuum* 26, no. 5 (2012): 669–683, doi:10.1080/10304312.2012.706457.

40. Jo Tacchi, "Open Content Creation: The Issues of Voice and the Challenges of Listening," *New Media & Society* 14, no. 4 (November 17, 2011): 652–668, doi:10.1177/1461444811422431; Jo Tacchi, Jerry Watkins, and Kosala Keerthirathne, "Participatory Content Creation: Voice, Communication, and Development," *Development in Practice* 19, no. 4–5 (June 2009): 573–584, doi:10.1080/09614520902866389.

41. Gerard Goggin and Jacqueline Clark, "Mobile Phones and Community Development: A Contact Zone between Media and Citizenship," *Development in Practice* 19, no. 4–5 (June 14, 2009): 585–597, doi:10.1080/09614520902866371.

42. Åke Grönlund, Rebekah Heacock, David Sasaki, Johan Hellström, and Walid Al-Saqaf, *Increasing Transparency & Fighting Corruption through ICT: Empowering People & Communities*, *SPIDER ICTD Series #3* (Stockholm: Universitetttervice US-AB, 2010).

43. François Bar, Melissa Brough, S. Costanza-Chock, Carmen Gonzalez, Cara J. Wallis, and Amanda Garces, "Mobile Voices: A Mobile, Open Source, Popular Communication Platform for First-Generation Immigrants in Los Angeles," in *"Mobile 2.0: Beyond Voice?" Pre-Conference Workshop at the International Communication Association (ICA)* (Chicago, 2009), http://lirneasia.net/wp-content/uploads/2009/05/final-paper_bar_et_al.pdf.

44. Heather Leson, "Series: Monitoring Corruption in Macedonia," *Ushahidi Blog*, June 20, 2013, http://www.ushahidi.com/2013/06/20/series-monitoring-corruption-in-macedonia/.

45. Ndesanjo Macha, "Geo-Mapping Tools and Dat-Analysis Redefine Reporting in-Africa," *Global Voices*, March 31, 2014, http://globalvoicesonline.org/2014/03/31/geo-mapping-tools-and-data-analysis-redefine-reporting-in-africa/.

46. Tavis Lupick, "Africa's Digital Election Trackers," *Al Jazeera*, December 12, 2012, http://www.aljazeera.com/indepth/features/2012/12/201212101152794146.html.

47. Bev Clark and Brenda Burrell, "Freedom Fone: Dial-up Information Service," in *ICTD'09 Proceedings of the 3rd International Conference on Information and Communication Technologies and Development* (Washington, DC: IEEE, 2009), 483, doi:10.1109/ICTD.2009.5426724.

48. Aditya Vashistha and William Thies, "IVR Junction: Building Scalable and Distributed Voice Forums in the Developing World," 6th USENIX/ACM Workshop on Networked Systems for Developing Regions, June 15, 2012, https://www.usenix.org/system/files/conference/nsdr12/nsdr12-final4.pdf.

49. Manas Mittal, Wei Wu, Steve Rubin, Sam Madden, and Björn Hartmann, "Bribecaster," In *Proceedings of the ACM 2012 Conference on Computer Supported Cooperative Work Companion—CSCW '12* (New York: ACM Press, 2012), 171–174, doi:10.1145/2141512.2141570.

50. Janaagraha, "Asked for a Bribe? Stay Calm and Report It," ipaidabribe.com, February 25, 2014, http://www.ipaidabribe.com/mobile.

51. Mills et al., "MoJo in Action."

52. Johan Hellström and Brooke Bocast, "Many 'Likers' Do Not Constitute A Crowd: The Case Of Uganda's Not in My Country," in *ICT for Anti-Corruption, Democracy and Education in East Africa*, ed. Katja Sarajeva (Stockholm: Universitetsservice US-AB, 2013), 27–36.

53. M. Bratton, "Briefing: Citizens and Cell Phones in Africa," *African Affairs* 112, no. 447 (February 7, 2013): 304–319, doi:10.1093/afraf/adt004; Thomas N. Smyth and Michael L. Best, "Tweet to Trust," in *Proceedings of the Sixth International Conference on Information and Communication Technologies and Development Full Papers—ICTD '13*, vol. 1 (New York: ACM Press, 2013), 133–141, doi:10.1145/2516604.2516617.

54. Jidraph Njuguna, Darshan Santani, Tierra Bills, Aisha W. Bryant, and Reginald Bryant, "Citizen Engagement and Awareness of the Road Surface Conditions in Nairobi, Kenya," in *Proceedings of the Fifth ACM Symposium on Computing for Development—ACM DEV-5 '14* (New York: ACM Press, 2014), 115–116, doi:10.1145/2674377.2678267.

55. Michael Champanis and Ulrike Rivett, "Reporting Water Quality," in *Proceedings of the Fifth International Conference on Information and Communication Technologies and Development—ICTD '12* (New York: ACM Press, 2012), 105–113, doi:10.1145/2160673.2160688.

56. Jennifer Whittal, "The Potential Use of Cellular Phone Technology in Maintaining an Up-to-Date Register of Land Transactions for the Urban Poor," *Potchefstroomse Elektroniese Regsblad* 14, no. 3 (2011), 162–194.

57. My colleagues at Microsoft Research India were looking at location-based approaches to monitoring teacher attendance. Azarias Reda, Saurabh Panjwani, and Edward Cutrell, "Hyke," in *Proceedings of the 5th ACM Workshop on Networked Systems for Developing Regions—NSDR '11* (New York: ACM Press, 2011), 15–20, doi:10.1145/1999927.1999933.

58. Antina Von Schnitzler, "Citizenship Prepaid: Water, Calculability, and Techno-Politics in South Africa," *Journal of Southern African Studies* 34, no. 4 (2008): 899–917, doi:10.1080/03057070802456821.

59. Jurgenson, "When Atoms Meet Bits."

60. Hellström and Bocast, "Many 'Likers' Do Not Constitute A Crowd."

61. Fabien Miard, "Call for Power? Mobile Phones as Facilitators of Political Activism," in *Cyberspaces and Global Affairs*, ed. Sean S. Costigan and Jake Perry (Aldershot, UK: Ashgate, 2012), 119–144; Muzammil M. Hussain and Philip N. Howard, "What Best Explains Successful Protest Cascades? ICTs and the Fuzzy Causes of the Arab Spring," *International Studies Review* 15, no. 1 (March 10, 2013): 48–66, doi:10.1111/misr.12020; Evgeny Morozov, *The Net Delusion: The Dark Side of Internet Freedom* (New York: Public Affairs, 2011).

62. John L. Sullivan, "Uncovering the Data Panopticon: The Urgent Need for Critical Scholarship in an Era of Corporate and Government Surveillance," *The Political Economy of Communication* 1, no. 2 (2013): np.

63. Jon Russell, "Thailand's Government Claims It Can Monitor the Country's 30M Line Users," *TechCrunch*, December 23, 2014, http://techcrunch.com/2014/12/23/thailand-line-monitoring-claim/.

64. Jan H. Pierskalla and Florian M. Hollenbach, "Technology and Collective Action: The Effect of Cell Phone Coverage on Political Violence in Africa," *American Political*

Science Review 107, no. 2 (March 27, 2013): 207–224, doi:http://dx.doi.org/10.1017/S0003055413000075.

65. Jon Boone, "Taliban Target Mobile Phone Masts to Prevent Tipoffs from Afghan Civilians," *GuardianUK Online*, November 11, 2011, http://www.theguardian.com/world/2011/nov/11/taliban-targets-mobile-phone-masts; Staff writer, "Boko Haram Destroys 25 Masts, Reportedly Kill 4 MTN Staff," *Balancing Act Africa*, September 7, 2012, http://www.balancingact-africa.com/news/en/issue-no-621/telecoms/boko-haram-destroys/en.

66. Yasacan Mustafa, "Living with Taksim Gezi: A Photo Essay," in *Reflections on Taksim—Gezi Park Protests in Turkey*, ed. Bülent Gökay and Ilia Xypolia (Staffs, UK: Keele University, 2013), 7.

67. Ilya Mouzykantskii, "Phone Charging Station #euromaidan," Twitpic, n.d., http://twitpic.com/douqbf.

68. Bhaveer Bhana, Stephen Flowerday, and Aharon Satt, "Using Participatory Crowdsourcing in South Africa to Create a Safer Living Environment," *International Journal of Distributed Sensor Networks* 2013 (2013): 1–13, doi:10.1155/2013/907196.

69. Guillaume Cornu Sébastien Le Bel, David Chavernac, and George Mapuvire, "FrontlineSMS as An Early Warning Network for Human-Wildlife Mitigation: Lessons Learned from Tests Conducted in Mozambique and Zimbabwe," *Electronic Journal of Information Systems in Developing Countries* 60 (2014), http://www.ejisdc.org/ojs2/index.php/ejisdc/article/view/1256.

70. Sarah Elwood, Michael F. Goodchild, and Daniel Z. Sui, "Researching Volunteered Geographic Information: Spatial Data, Geographic Research, and New Social Practice," *Annals of the Association of American Geographers* 102, no. 3 (May 2012): 571–590, doi:10.1080/00045608.2011.595657; Mark H. Palmer and Scott Kraushaar, eds., *Crowdsourcing Geographic Knowledge: Volunteered Geographic Information (VGI) in Theory and Practice* (Amsterdam: Springer Netherlands, 2013), doi:10.1007/978-94-007-4587-2.

71. Mark Graham and Håvard Haarstad, "Open Development through Open Consumption: The Internet of Things, User-Generated Content and Economic Transparency," in *Open Development*, ed. Matthew L. Smith and Katherine M. A. Reilly (Cambridge, MA: MIT Press, 2013), 79–111.

72. Halewood and Surya, "Mobilizing the Agricultural Value Chain."

73. Daniel Soto, Edwin Adkins, Matt Basinger, Rajesh Menon, Sebastian Rodriguez-Sanchez, Natasha Owczarek, Ivan Willig, and Vijay Modi, "A Prepaid Architecture for Solar Electricity Delivery in Rural Areas," in *Proceedings of the Fifth International Conference on Information and Communication Technologies and Development—ICTD '12* (New York: ACM Press, 2012), 130–138, doi:10.1145/2160673.2160691.

74. Ragnhild Overå, "Networks, Distance, and Trust: Telecommunications Development and Changing Trading Practices in Ghana," *World Development* 34, no. 7 (July 2006): 1301–1315, doi:doi:10.1016/j.worlddev.2005.11.015.

75. Erik Hersman, "Sendy: Digitizing Motorcycle Deliveries," Whiteafrican.com, September 23, 2014, http://whiteafrican.com/2014/09/23/sendy-digitizing -motorcycle-deliveries/.

76. James Krohe, Jr., "Not Your Daddy's Taxi," *Planning* 79, no. 5 (2013): 15–17; Arun Kumar, Nitendra Rajput, Sheetal Agarwal, Dipanjan Chakraborty, and Amit Anit Nanavati, "Organizing the Unorganized—Employing IT to Empower the Under-Privileged," in *Proceedings of the 17th International Conference on World Wide Web—WWW '08* (New York: ACM Press, 2008), 935–944, doi:10.1145/1367497 .1367623.

77. Rohan Samarajiva, "Policy Commentary: Mobilizing Information and Communications Technologies for Effective Disaster Warning: Lessons from the 2004 Tsunami," *New Media & Society* 7, no. 6 (December 2005): 731–747, doi:10.1177/ 1461444805058159.

78. Alette Schoon, "Dragging Young People down the Drain: The Mobile Phone, Gossip Mobile Website Outoilet and the Creation of a Mobile Ghetto," *Critical Arts* 26, no. 5 (November 2012): 690–706, doi:10.1080/02560046.2012.744723; Fie Velghe, "Deprivation, Distance and Connectivity: The Adaptation of Mobile Phone Use to a Life in Wesbank, a Post-Apartheid Township in South Africa," *Discourse, Context & Media* 1, no. 4 (2011): 203–216, doi:http://dx.doi.org/10.1016/j.dcm .2012.09.004.

79. Assa Doron and Robin Jeffrey, *The Great Indian Phone Book: How the Cheap Cell Phone Changes Business, Politics, and Daily Life* (Cambridge, MA: Harvard University Press, 2013).

80. Pierskalla and Hollenbach, "Technology and Collective Action."

81. Cara J. Wallis, "Mobile Phones without Guarantees: The Promises of Technology and the Contingencies of Culture," *New Media & Society* 13, no. 3 (March 25, 2011): 471–485, doi:10.1177/1461444810393904; Jack Linchuan Qiu, "The Wireless Leash: Mobile Messaging Service as a Means of Control," *International Journal of Communication* 1, no. 1 (2007): 74–91.

82. Ithiel de Sola Pool, "Introduction," *The Social Impact of the Telephone*, ed. Ithiel de Sola Pool (Cambridge, MA: MIT Press, 1977), 1–9, p. 4.

Chapter 8

1. Michael Gurstein, *Community Informatics: Enabling Communities with Information and Communications Technologies* (Hershey, PA: Idea Group, 2000).

2. Mirca Madianou and Daniel Miller, "Polymedia: Towards a New Theory of Digital Media in Interpersonal Communication," *International Journal of Cultural Studies* 16, no. 2 (August 22, 2012): 169–187, doi:10.1177/1367877912452486.

3. Nancy Baym, *Personal Connections in the Digital Age* (Cambridge, UK: Polity, 2010); Stefana Broadbent and Valerie Bauwens, "Understanding Convergence," *Interactions* 15, no. 1 (2008): 23–27; Nick Couldry, *Media, Society, World: Social Theory and Digital Media Practice* (Cambridge, UK: Polity, 2012); Henry Jenkins, *Convergence Culture: Where Old and New Media Collide* (New York: New York University Press, 2006).

4. Madianou and Miller, "Polymedia," 176.

5. Thorsten Quandt and Thilo Von Pape, "Living in the Mediatope: A Multimethod Study on the Evolution of Media Technologies in the Domestic Environment," in *Understanding Creative Users of ICTs: Users as Social Actors*, ed. David Kurt Herold, Harmeet Sawhney, and Leopoldina Fortunati (New York: Routledge, 2013), 83–98.

6. Dorothea Kleine, *Technologies of Choice? ICTs, Development, and the Capabilities Approach* (Cambridge, MA: MIT Press, 2013).

7. Jandy Luik, "The Importance of Fluidity Utility Belief and Technology Cluster Ownership on Adoption of Mobile Communication among Youth," *Proceedings of the 22nd Annual Conference of the Asian Media Information and Communication Centre* (Yogyakarta, Indonesia, June 2013), http://repository.petra.ac.id/16139/; Harmeet Sawhney, "Strategies for Increasing the Conceptual Yield of New Technologies Research," *Communication Monographs* 74, no. 3 (September 2007): 395–401, doi:10.1080/03637750701543527.

8. Arnau Monterde and John Postill, "Mobile Ensembles: The Uses of Mobile Phones for Social Protest by Spain's Indignados," *Routledge Companion to Mobile Media*, ed. Gerard Goggin and Larissa Hjorth (New York: Routledge, 2014), 429–438.

9. Melissa Densmore, Ben Bellows, John Chuang, and Eric Brewer, "The Evolving Braid," in *Proceedings of the Sixth International Conference on Information and Communication Technologies and Development Full Papers—ICTD '13*, vol. 1 (New York: ACM Press, 2013), 257–266, doi:10.1145/2516604.2516620.

10. Ralph Schroeder, "Mobile Phones and the Inexorable Advance of Multimodal Connectedness," *New Media & Society* 12, no. 1 (February 19, 2010): 75–90, doi:10.1177/1461444809355114.

11. John Christian Feaster, "The Repertoire Niches of Interpersonal Media: Competition and Coexistence at the Level of the Individual," *New Media & Society* 11, no. 6 (September 7, 2009): 965–984, doi:10.1177/1461444809336549; Harsh Taneja, James G. Webster, Edward C. Malthouse, and Thomas B. Ksiazek, "Media Consumption across Platforms: Identifying User-Defined Repertoires," *New Media & Society* 14, no. 6 (March 20, 2012): 951–968, doi:10.1177/1461444811436146.

12. Rocío Gómez, "The Ecology of Linking Technologies: Toward a Non-Instrumental Look at New Technological Repertoires," *The Journal of Community Informatics* 9, no. 3 (2013): np.

13. Nigel Thrift, "New Urban Eras and Old Technological Fears: Reconfiguring the Goodwill of Electronic Things," *Urban Studies* 33, no. 8 (October 1, 1996): 1463–1494, doi:10.1080/0042098966754.

14. Cornelia Wolf and Anna Schnauber, "The Information Repertoire of Mobile Internet Users: First Hints for the Digital Transition?," paper presented at the 2013 ICA Preconference on Mobile Communication, *10 Years On: Looking Forward in Mobile ICT Research,* May 17, 2013, London; Joey Reagan, "The 'Repertoire' of Information Sources," *Journal of Broadcasting & Electronic Media* 40, no. 1 (January 1996): 112–121, doi:10.1080/08838159609364336.

15. Wanda J. Orlikowski and JoAnne Yates, "Genre Repertoire: The Structuring of Communicative Practices in Organizations," *Administrative Science Quarterly* 39, no. 4 (1994): 541–574.

16. Andrew Chadwick, "Digital Network Repertoires and Organizational Hybridity," *Political Communication* 24, no. 3 (August 6, 2007): 283–301, doi:10.1080/10584600701471666.

17. Paulene E. W. van den Berg, Theo A. Arentze, and Harry J. P. Timmermans, "New ICTs and Social Interaction: Modelling Communication Frequency and Communication Mode Choice," *New Media & Society* 14, no. 6 (March 21, 2012): 987–1003, doi:10.1177/1461444812437518.

18. Bonnie A. Nardi and Vicki L. O'Day, *Information Ecologies: Using Technology with Heart* (Cambridge, MA: MIT Press, 1999).

19. Richard Heeks, "Information and Communication Technologies, Poverty and Development," *Institute for Development Policy and Management Working Paper Series* #5 (Manchester, UK, June 1999), http://www.sed.manchester.ac.uk/idpm/research/publications/wp/di/documents/di_wp05.pdf.

20. Nardi and O'Day, *Information Ecologies.*

21. Bill Maurer, "Mobile Money: Communication, Consumption and Change in the Payments Space," *Journal of Development Studies* 48, no. 5 (2012): 1–16, doi:10.1080/00220388.2011.621944.

22. Erin B. Taylor and Heather A. Horst, "The Aesthetics of Mobile Money Platforms in Haiti," in *Routledge Companion to Mobile Media*, ed. Gerard Goggin and Larissa Hjorth (London: Routledge, 2014), 462–471, p. 462.

23. Raul Pertierra, "Localizing Mobile Media: A Philippine Perspective," in *Routledge Companion to Mobile Media*, ed. Gerard Goggin and Larissa Hjorth (New York: Rout-

ledge, 2014), 51. See also Nic Bidwell, "Ubuntu in the Network: Humanness in Social Capital in Rural Africa," *Interactions* 17, no. 2 (March 1, 2010): 68–71, doi:10.1145/1699775.1699791.

24. Don Slater, *New Media, Development & Globalization* (Cambridge, UK: Polity Press, 2013), 27–67.

25. Wallis has used the assemblage lens to great effect in her analysis of ICT use in China; see Cara Wallis, *Technomobility in China: Young Migrant Women and Mobile Phones* (New York: New York University Press, 2013).

26. Slater, *New Media, Development & Globalization*, 48.

27. Mark Rodini, Michael R. Ward, and Glenn A. Woroch, "Going Mobile: Substitutability between Fixed and Mobile Access," *Telecommunications Policy* 27, no. 5–6 (June 2003): 457–476, doi:10.1016/S0308-5961(03)00010-7.

28. Chris Taubman, Maria Vagliasindi, and Izzet Guney, "Fixed and Mobile Competition in Transition Economies," *Telecommunications Policy* 30, no. 7 (August 2006): 349–367, doi:10.1016/j.telpol.2006.02.002; Jacqueline Hamilton, "Are Main Lines and Mobile Phones Substitutes or Complements? Evidence from Africa," *Telecommunications Policy* 27, no. 1–2 (2003): 109–133, doi:10.1016/S0308-5961(02)00089-7.

29. Leslie Haddon and Jane Vincent, "Making the Most of the Communications Repertoire: Choosing between the Mobile and Fixed-Line," in *The Global and the Local in Mobile Communication*, ed. Kristóf Nyíri (Vienna: Passagen Verlag, 2005), 231–240; Christian Licoppe, "'Connected' Presence: The Emergence of a New Repertoire for Managing Social Relationships in a Changing Communication Technoscape," *Environment and Planning D: Society and Space* 22, no. 1 (2004): 135–156, doi:10.1068/d323t.

30. Katy E. Pearce, "Personal Communication," August 2014.

31. Joo-Young Jung, Wan-Ying Lin, and Yong-Chan Kim, "The Dynamic Relationship between East Asian Adolescents' Use of the Internet and Their Use of Other Media," *New Media & Society* 14, no. 6 (March 15, 2012): 969–986, doi:10.1177/1461444812437516.

32. Petter Nielsen and Annita Fjuk, "The Reality beyond the Hype: Mobile Internet Is Primarily an Extension of PC-Based Internet," *The Information Society* 26, no. 5 (October 27, 2010): 375–382, doi:10.1080/01972243.2010.511561. See also Mark de Reuver, Guido Ongena, and Harry Bouwman, "Should Mobile Internet Be an Extension to the Fixed Web? Fixed-Mobile Reinforcement as Mediator between Context of Use and Future Use," *Telematics and Informatics* 30, no. 2 (May 2013): 111–120, doi:10.1016/j.tele.2012.02.002.

33. Wan-Ying Lin, Xinzhi Zhang, Joo-Young Jung, and Yong-Chan Kim, "From the Wired to Wireless Generation? Investigating Teens' Internet Use through the

Mobile Phone," *Telecommunications Policy* 37, no. 8 (September 2013): 651–661, doi:10.1016/j.telpol.2012.09.008.

34. M. Cristina Ciancetta, Giovanni Colombo, Raffaella Lavagnolo, and Davide Grillo, "Convergence Trends for Fixed and Mobile Services," *IEEE Personal Communications* 6, no. 2 (April 1, 1999): 14–21, doi:10.1109/98.760419.

35. Pauk Fox, Jonathan Wareham, Davar Rezania, and Ellen Christiaanse, "Will Mobiles Dream of Electric Sheep? Expectations of the New Generation of Mobile Users: Misfits with Practice and Research," in *ICMB '06 Proceedings of the International Conference on Mobile Business* (Washington, DC: IEEE Computer Society, 2006), 44, doi:10.1109/ICMB.2006.50.

36. Helani Galpaya, Rohan Samarajiva, and Shamistra Soysa, "Taking E-Government to the Bottom of the Pyramid," in *Proceedings of the 1st International Conference on Theory and Practice of Electronic Governance—ICEGOV '07* (New York: ACM Press, 2007), 233, doi:10.1145/1328057.1328105. See also Ayesha Zainudeen, "Are the Poor Stuck in Voice? Conditions for Adoption of More-Than-Voice Mobile Services," *Information Technologies & International Development* 7, no. 3 (2011): 45–59.

37. Jonathan Donner, Shikoh Gitau, and Gary Marsden, "Exploring Mobile-Only Internet Use: Results of a Training Study in Urban South Africa," *International Journal of Communication* 5 (2011): 574–597; Jonathan Donner and Shikoh Gitau, "New Paths: Exploring Mobile-Centric Internet Use in South Africa," "Mobile 2.0: Beyond Voice?" Pre-Conference Workshop at the International Communication Association (ICA) (Chicago, April 2009), http://lirneasia.net/wp-content/uploads/2009/05/final-paper_donner_et_al.pdf; Shikoh Gitau, Gary Marsden, and Jonathan Donner, "After Access: Challenges Facing Mobile-Only Internet Users in the Developing World," in *Proceedings of the 28th International Conference on Human Factors in Computing Systems—CHI '10* (New York: ACM, 2010), 2603–2606, doi:10.1145/1753326.1753720.

38. LIRNEasia, "Mobile 2.0: Beyond Voice? ICA Preconference Program," May 2009, http://lirneasia.net/wp-content/uploads/2009/05/mobile20-preconference-program_revised.pdf.

39. Jonathan Donner and Marion Walton, "Your Phone Has Internet—Why Are You at a Library PC? Re-Imagining Public Access for the Mobile Internet Era," in *Proceedings of INTERACT 2013: 14th IFIP TC 13 International Conference* (Berlin: Springer, 2013), 347–364, doi:10.1007/978-3-642-40483-2_25.

40. François Bar, Chris Coward, Lucas Koepke, Chris Rothschild, Araba Sey, and George Sciadas, "The Impact of Public Access to ICTs," in *Proceedings of the Sixth International Conference on Information and Communication Technologies and Development Full Papers—ICTD '13*, vol. 1 (New York: ACM Press, 2013), 34–42, doi:10.1145/2516604.2516619.

41. Mastin Prinsloo and Marion Walton, "Situated Responses to the Digital Literacies of Electronic Communication in Marginal School Settings," in *Yearbook 2008: African Media, African Children*, ed. Norma Pecora, Enyonam Osei-Hwere, and Ulla Carlson (Göteborg: Nordicom, Göteborgs Universitet, 2008), 99–116.

42. Gary Marsden, "What Is the Mobile Internet?," *Interactions* 14, no. 6 (November 2007): 24–25, doi:10.1145/1300655.1300672.

43. Wallace Chigona, Guy Kankwenda, and Saffia Manjoo, "The Uses and Gratifications of Mobile Internet among the South African Students," in *PICMET '08–2008 Portland International Conference on Management of Engineering & Technology (Cape Town July 27–31)*, vol. 10 (Washington, DC: IEEE, 2008), 2197–2207, doi:10.1109/PICMET.2008.4599842; Wallace Chigona, Darry Beukes, Junaid Vally, and Maureen Tanner, "Can Mobile Internet Help Alleviate Social Exclusion in Developing Countries?," *The Electronic Journal of Information Systems in Developing Countries* 36 (2009): 1–16.

44. Merryl Ford and Adele Botha, "MobilED–An Accessible Mobile Learning Platform for Africa," in *IST-Africa 2007 Conference Proceedings*, ed. Paul Cunningham and Miriam Cunningham (IIMC International Information Management Corporation, 2007), 9–11.

45. Tanja Bosch, "Wots Ur ASLR? Adolescent Girls' Use of Cellphones in Cape Town," *Commonwealth Youth and Development* 6, no. 2 (2008): 52–69.

46. Katy E. Pearce and Ronald E. Rice, "Digital Divides from Access to Activities: Comparing Mobile and Personal Computer Internet Users," *Journal of Communication* 63, no. 4 (2013): 721–744, doi:10.1111/jcom.12045; Katy E. Pearce, Janine Slaker, and Nida Ahmad, "Is Your Web Everyone's Web? Theorizing the Web through the Lens of the Device Divide," *Theorizing the Web 2012*, April 12, 2012, http://fr.slideshare.net/katyp1/katy-pearce-ttw12; Katy E. Pearce, "Phoning It In: Theory in Mobile Media and Communication in Developing Countries," *Mobile Media & Communication* 1, no. 1 (January 1, 2013): 76–82, doi:10.1177/2050157912459182.

47. Neha Kumar, "Facebook for Self-Empowerment? A Study of Facebook Adoption in Urban India," *New Media & Society*, July 25, 2014, doi:10.1177/1461444814543999.

48. Jerry Watkins, Kathi R. Kitner, and Dina Mehta, "Mobile and Smartphone Use in Urban and Rural India," *Continuum* 26, no. 5 (October 2012): 685–697, doi:10.1080/10304312.2012.706458.

49. Nimmi Rangaswamy and Edward Cutrell, "Anthropology, Development and ICTs," in *Proceedings of the Fifth International Conference on Information and Communication Technologies and Development—ICTD '12* (New York: ACM Press, 2012), 85–93, doi:10.1145/2160673.2160685; Nimmi Rangaswamy and S. Yamsani, "'Mental Kartha Hai' or 'Its Blowing My Mind': Evolution of the Mobile Internet in an Indian

Slum," in *Proceedings of the 2001 Ethnographic Praxis in Industry Conference* (American Anthropological Association, 2011), 285–298, https://www.epicpeople.org/wp-content/uploads/2014/09/Rangaswamy_menta.pdf; Preeti Mudliar and Nimmi Rangaswamy, "Offline Strangers, Online Friends," in *Proceedings of the 33rd Annual ACM Conference on Human Factors in Computing Systems—CHI '15* (New York: ACM, 2015), 3799–3808, doi:10.1145/2702123.2702533.

50. Maeve Duggan and Aaron Smith, "Cell Internet Use 2013," *Pew Research Center Reports*, September 16, 2013, http://www.pewinternet.org/files/old-media//Files/Reports/2013/PIP_CellInternetUse2013.pdf.

51. Lynn Schofield Clark, Christof Demont-Heinrich, and Scott Webber, "Parents, ICTs, and Children's Prospects for Success: Interviews along the Digital 'Access Rainbow,'" *Critical Studies in Media Communication* 22, no. 5 (2005): 409–426, doi:10.1080/07393180500342985.

52. Sakari Taipale, "Do the Mobile-Rich Get Richer? Internet Use, Travelling and Social Differentiations in Finland," *New Media & Society*, published online before print (May 26, 2014), doi:10.1177/1461444814536574.

53. Darja Groselj and Grant Blank, "Comparing Mobile and Non-Mobile Internet Users: How Are Mobile Users Different?," paper presented at the 2013 ICA Preconference on Mobile Communication, *10 Years On: Looking Forward in Mobile ICT Research,* May 17, 2013, London.

54. Lee Humphreys, Thilo Von Pape, and Veronika Karnowski, "Evolving Mobile Media: Uses and Conceptualizations of the Mobile Internet," *Journal of Computer-Mediated Communication* 18, no. 4 (July 30, 2013): 491–507, doi:10.1111/jcc4.12019, p. 496.

55. Eunmo Sung and Richard E. Mayer, "Students' Beliefs about Mobile Devices vs. Desktop Computers in South Korea and the United States," *Computers & Education* 59, no. 4 (December 2012): 1328–1338, doi:10.1016/j.compedu.2012.05.005.

56. Gerard Goggin, "Driving the Internet: Mobile Internets, Cars, and the Social," *Future Internet* 4, no. 4 (March 20, 2012): 306–321, doi:10.3390/fi4010306, p. 307.

57. Pew Research Center, "Emerging Nations Embrace Internet, Mobile Technology: Cell Phones Nearly Ubiquitous in Many Countries," February 2014, http://www.pewglobal.org/files/2014/02/Pew-Research-Center-Global-Attitudes-Project-Technology-Report-FINAL-February-13-20146.pdf.

58. Drew DeSilver, "Overseas Users Power Facebook's Growth; More Going Mobile-Only," *Pew Research Fact-Tank*, February 4, 2013, http://www.pewresearch.org/fact-tank/2014/02/04/overseas-users-power-facebooks-growth-more-going-mobile-only/.

59. Alex Williams, "5 Perspectives on the Future of the Human Interface," *Techcrunch*, November 4, 2012, http://techcrunch.com/2012/11/04/5-perspectives -on-the-future-of-the-human-interface/.

60. Cory Doctorow, "Crowdfunding an Ubuntu Phone That Doubles as Your PC," *BoingBoing*, July 22, 2013, http://boingboing.net/2013/07/22/crowdfunding-an-ubuntu-phone-t.html.

61. MG Siegler, "'It's Just A Big iPod Touch,'" *Techcrunch*, November 11, 2013, http://techcrunch.com/2013/11/11/ipad-air-my-mom-keyboards/.

62. Jessica Guynn, "Who's on Mobile First? Google, Facebook Race to Get There First," *The Los Angeles Times*, October 12, 2012, http://articles.latimes.com/2012/ oct/12/business/la-fi-tn-google-facebook-mobile-first-20121012; Ed Hardy, "Google Adopts a New Strategy: Mobile First," *Brighthand.com*, February 17, 2010, http:// www.brighthand.com/default.asp?newsID=16235&news=Google+Android+OS +CEO+Eric+Schmidt+Mobile+First.

63. Andrea Kavanaugh, Anita Puckett, and Deborah Tatar, "Scaffolding Technology for Low Literacy Groups: From Mobile Phone to Desktop PC?," *International Journal of Human–Computer Interaction* 29, no. 4 (March 2013): 274–288, doi:10.1080/ 10447318.2013.765766.

64. Henry Blodget, "Anyone Who Believes in 'Mobile First' Needs to Look at This Photo," *Business Insider*, December 14, 2012, http://www.businessinsider.com/ mobile-first-not-2012-12.

65. MG Siegler, "My Product Feedback," Massivegreatness.com, September 20, 2012, http://massivegreatness.com/mobile.

66. Eszter Hargittai and Gina Walejko, "The Participation Divide: Content Creation and Sharing in the Digital Age," *Information, Communication & Society* 11, no. 2 (March 2008): 239–256, doi:10.1080/13691180801946150; Eszter Hargittai and Yuli Patrick Hsieh, "Digital Inequality," in *The Oxford Handbook of Internet Studies*, ed. William H. Dutton (Oxford, UK: Oxford University Press, 2013), 129–150; Eszter Hargittai, "Second-Level Digital Divide: Differences in People's Online Skills," *First Monday* 7, no. 4 (2002): 1–20; Eszter Hargittai, "Digital Na(t)ives? Variation in Internet Skills and Uses among Members of the 'Net Generation,'" *Sociological Inquiry* 80, no. 1 (2010): 92–113, doi:10.1111/j.1475-682X.2009.00317.x.

67. Julie Coiro, Michele Knobel, Colin Lankshear, and Donald J. Leu, eds., *Handbook of Research on New Literacies* (New York: Routledge, 2008); Sonia Livingstone, "Media Literacy and the Challenge of New Information and Communication Technologies," *The Communication Review* 7, no. 1 (January 2004): 3–14, doi:10.1080/ 10714420490280152; Jean A. Wooldridge, "Digital Literacy in a Landscape of Data: A Plea for a Broader Definition for Citizens and Patients.," *Studies in Health Technology and Informatics* 118 (January 2005): 263–269.

68. Alexander J. A. M. van Deursen, Jan van Dijk, and Oscar Peters, "Rethinking Internet Skills: The Contribution of Gender, Age, Education, Internet Experience, and Hours Online to Medium- and Content-Related Internet Skills," *Poetics* 39, no. 2 (April 2011): 125–144, doi:10.1016/j.poetic.2011.02.001; Eden Litt, "Measuring Users' Internet Skills: A Review of Past Assessments and a Look toward the Future," *New Media & Society* 15, no. 4 (May 24, 2013): 612–630, doi:10.1177/1461444813475424.

69. Steven Sam, "Exploring Mobile Internet Use among Marginalised Young People in Post-Conflict Sierra Leone," *The Electronic Journal of Information Systems in Developing Countries* 66, no. 5 (2015): 1–20.

70. Dinuka Wijetunga, "The Digital Divide Objectified in the Design: Use of the Mobile Telephone by Underprivileged Youth in Sri Lanka," *Journal of Computer-Mediated Communication* 19, no. 3 (2014): 712–726, doi:10.1111/jcc4.12071.

71. Hassan Baig, "Peering into the Minds of the 4.3 Billion Unconnected," *Techcrunch*, November 29, 2014, http://techcrunch.com/2014/11/29/peering-into -the-minds-of-the-4-3-billion-unconnected/?ncid=rss.

72. Maria Rosa Lorini, Izak van Zyl, and Wallace Chigona, "ICTs for Inclusive Communities: A Critical Discourse Analysis," in *Proceedings of the 8th International Development Informatics Association Conference—IDIA 2014* (Melborne, Victoria: Monash University, 2014), np, www.developmentinformatics.org/conferences/2014/papers/7-Lorini-VanZyl-Chigona.pdf; Gitau, Marsden, and Donner, "After Access."

73. Yong Jin Park, "My Whole World's in My Palm! The Second-Level Divide of Teenagers' Mobile Use and Skill," *New Media & Society*, published online before print, January 27, 2014, doi:10.1177/1461444813520302.

74. Eszter Hargittai, "Second-Level Digital Divide: Differences in People's Online Skills," *First Monday* 7, no. 4 (2002), np, http://firstmonday.org/article/view/942/864; Eszter Hargittai and Aaron Shaw, "Mind the Skills Gap: The Role of Internet Know-How and Gender in Differentiated Contributions to Wikipedia," *Information, Communication & Society* 18, no. 4 (2015): 424–442, doi:10.1080/1369118X.2014.957711; Nicole Zillien and Eszter Hargittai, "Digital Distinction: Status-Specific Types of Internet Usage," *Social Science Quarterly* 90, no. 2 (June 2009): 274–291, doi:10.1111/j.1540-6237.2009.00617.x; Hargittai and Hsieh, "Digital Inequality."

75. Indrani Medhi, Somani Patnaik, Emma Brunskill, S. N. Nagasena Gautama, William Thies, and Kentaro Toyama, "Designing Mobile Interfaces for Novice and Low-Literacy Users," *ACM Transactions on Computer-Human Interaction* 18, no. 1 (2011): 1–28, doi:10.1145/1959022.1959024.

76. Joyojeet Pal, Tawfiq Ammari, Ramaswami Mahalingam, Ana Maria Huaita Alfaro, and Meera Lakshmanan, "Marginality, Aspiration and Accessibility in ICTD," in *Proceedings of the Sixth International Conference on Information and Communication*

Technologies and Development Full Papers—ICTD '13, vol. 1 (New York: ACM Press, 2013), 68–78, doi:10.1145/2516604.2516623.

77. Brenda Danet and Susan C. Herring, eds., *The Multilingual Internet* (Oxford, UK: Oxford University Press, 2007); Katy E. Pearce and Ronald E. Rice, "The Language Divide—The Persistence of English Proficiency as a Gateway to the Internet: The Cases of Armenia, Azerbaijan, and Georgia," *International Journal of Communication* 8 (2014): 2834–2859; Iris Orriss, "The Internet's Language Barrier," *Innovations: Technology, Governance, Globalization* 9, no. 3–4 (July 2014): 123–126, doi:10.1162/inov_a_00223.

78. Kieran Mervyn, Anoush Simon, and David K. Allen, "Digital Inclusion and Social Inclusion: A Tale of Two Cities," *Information, Communication & Society* 17, no. 9 (2014): 1086–1104, doi:10.1080/1369118X.2013.877952.

79. Mark Warschauer, *Technology and Social Inclusion: Rethinking the Digital Divide* (Cambridge, MA: MIT Press, 2003).

80. Ibid., 8.

81. Michael Gurstein, "Effective Use: A Community Informatics Strategy beyond the Digital Divide," *First Monday* 8, no. 12 (December 1, 2003), doi:10.5210/fm.v8i12.1107.

82. Ibid.

83. Donghee Yvette Wohn, Cliff Lampe, Rick Wash, Nicole Ellison, and Jessica Vitak, "The 'S' in Social Network Games: Initiating, Maintaining, and Enhancing Relationships," in *2011 44th Hawaii International Conference on System Sciences* (New York: IEEE, 2011), 1–10, doi:10.1109/HICSS.2011.400.

84. Beth E. Kolko and Cynthia Putnam, "Computer Games in the Developing World: The Value of Non-Instrumental Engagement with ICTs, or Taking Play Seriously," in *Proceedings of 2009 International Conference on Information and Communication Technologies and Development—ICTD2009* (New York: IEEE, 2009), 46–55, doi:10.1109/ICTD.2009.5426705.

85. Michael Chan, "Mobile Phones and the Good Life: Examining the Relationships among Mobile Use, Social Capital and Subjective Well-Being," *New Media & Society* 17, no. 1 (2015): 96–113, doi:10.1177/1461444813516836.

86. Nicole B. Ellison, Charles Steinfield, and Cliff Lampe, "Connection Strategies: Social Capital Implications of Facebook-Enabled Communication Practices," *New Media & Society* 13, no. 6 (2011): 873–892, doi:10.1177/1461444810385389; Ellen Johanna Helsper, "A Corresponding Fields Model for the Links between Social and Digital Exclusion," *Communication Theory* 22, no. 4 (November 2012): 403–426, doi:10.1111/j.1468-2885.2012.01416.x; Cliff Lampe, Jessica Vitak, and Nicole B. Ellison, "Users and Nonusers: Interactions between Levels of Adoption and Social

Capital," in *Proceedings of the 2013 Conference on Computer Supported Cooperative Work—CSCW '13* (New York: ACM Press, 2013), 809–820, doi:10.1145/2441776 .2441867.

87. Amartya Sen, *Development as Freedom* (New York: Anchor Books, 2000).

88. Bridgette Wessels, "Exploring Human Agency and Digital Systems," *Information, Communication & Society* 16, no. 10 (December 2013): 1533–1552, doi:10.1080 /1369118X.2012.715666; Kleine, *Technologies of Choice?*; Matthew L. Smith, Kent Street, Randy Spence, and Ahmed T. Rashid, "Mobile Phones and Expanding Human Capabilities," *Information Technologies & International Development* 7, no. 3 (2011): 77–88.

89. Kleine, *Technologies of Choice?*, 9.

90. Marion Walton, "Pavement Internet: Mobile Media Economies and Ecologies in South Africa," in *Routledge Companion to Mobile Media*, ed. Gerard Goggin and Larissa Hjorth (London: Routledge, 2014), 450–461, p. 450.

91. Ibid., 451. See also Paul Leonardi, "Digital Materiality? How Artifacts without Matter, Matter," *First Monday* 15, no. 6 (2010), np, http://firstmonday.org/article/ view/3036/2567; Marion Walton and Pierrinne Leukes, "Prepaid Social Media and Mobile Discourse in South Africa," *Journal of African Media Studies* 5, no. 2 (2013): 149–167, doi:10.1386/jams.5.2.149_1.

92. Lisa Horner, *A Human Rights Approach to the Mobile Internet* (Mellville, South Africa: Association for Progressive Communications, 2011), http://www5.apc.org/es/ system/files/LisaHorner_MobileInternet-ONLINE.pdf, 13–14.

93. Philip M. Napoli and Jonathan A. Obar, "The Emerging Mobile Internet Underclass: A Critique of Mobile Internet Access," *The Information Society* 30, no. 5 (October 2, 2014): 323–334, doi:10.1080/01972243.2014.944726, p. 330.

94. Mark Surman, Corina Gardner, and David Ascher, "Local Content, Smartphones, and Digital Inclusion," *Innovations: Technology, Governance, Globalization* 9, no. 3–4 (July 2014): 63–74, doi:10.1162/inov_a_00217, p. 68.

95. Astra Taylor, *The People's Platform: Taking Back Power and Culture in the Digital Age* (New York: Metropolitan Books, 2014), 109. Susan Crawford, "The New Digital Divide," *The New York Times*, December 4, 2011.

96. International Telecommunication Union, *Measuring the Information Society, 2013* (Geneva, 2013).

97. Pádraig Carmody, "The Informationalization of Poverty in Africa? Mobile Phones and Economic Structure," *Information Technologies & International Development* 8, no. 3 (2012): 1–17; Chenxing Han, "South African Perspectives on Mobile Phones: Challenging the Optimistic Narrative of Mobiles for Development," *International Journal of Communication* 6 (2012): 2057–2081.

98. Pearce, Slaker, and Ahmad, "Is Your Web Everyone's Web? Theorizing the Web through the Lens of the Device Divide."

Chapter 9

1. Jonathan Donner and Marion Walton, "Your Phone Has Internet—Why Are You at a Library PC? Re-Imagining Public Access for the Mobile Internet Era," in *Proceedings of INTERACT 2013: 14th IFIP TC 13 International Conference* (Berlin: Springer, 2013), 347–364, doi:10.1007/978-3-642-40483-2_25.

2. Ibid., 354.

3. Kas Kalba, "The Adoption of Mobile Phones in Emerging Markets: Global Diffusion and the Rural Challenge," *International Journal of Communication* 2 (2008): 631–661.

4. Stuart Rutherford, *The Poor and Their Money* (Oxford, UK: Oxford University Press, 2001).

5. Marshini Chetty, Richard Banks, A. J. Bernheim Brush, Jonathan Donner, and Rebecca E. Grinter, "'You're Capped!' Understanding the Effects of Bandwidth Caps on Broadband Use in the Home," in *Proceedings of CHI 2012, May 5–10, 2010, Austin, TX* (New York: ACM, 2012), 3021–3030, doi:10.1145/2207676.2208714.

6. Calvin Taylor, "Pre-Paid Literacy: Negotiating the Cost of Adolescent Mobile Technology Use," *English in Australia* 44, no. 2 (2009): 26–34. See also Jane Duncan, "Mobile Network Society? Affordability and Mobile Phone Usage in Grahamstown East," *Communicatio: South African Journal for Communication Theory and Research* 39, no. 1 (March 2013): 35–52, doi:10.1080/02500167.2013.766224.

7. Regina De Angoitia and Fernando Ramirez, "Strategic Use of Mobile Telephony at the Bottom of the Pyramid: The Case of Mexico," *Information Technologies and International Development* 5, no. 3 (2009): 35–53.

8. Jonathan Donner, "The Rules of Beeping: Exchanging Messages Via Intentional 'Missed Calls' on Mobile Phones," *Journal of Computer-Mediated Communication* 13, no. 1 (October 2008): 1–22, doi:10.1111/j.1083-6101.2007.00383.x.

9. Ingrid Lunden, "Twitter Is in Talks with India's ZipDial," *TechCrunch*, January 12, 2015, http://techcrunch.com/2015/01/12/zipdial-twitter/.

10. Neilsen, "Pay-as-You-Phone: How Global Consumers Pay for Mobile," *Newswire*, 2013, http://www.nielsen.com/us/en/insights/news/2013/how-global-consumers -pay-for-mobile.html.

11. Leslie Haddon and Jane Vincent, "Children's Broadening Use of Mobile Phones," in *Mobile Technologies: From Telecommunications to Media*, ed. Gerard Goggin and Larissa Hjorth (New York: Routledge, 2009), 37–49.

12. Rich Ling and Päl Roe Sundsoy, "The iPhone and Mobile Access to the Internet," in *ICA Pre-Conference on Mobile Communication,* ed. LIRNEasia (Chicago: LIRNEasia, 2009).

13. Andrew Odlyzko, Bill St. Arnaud, Erik Stallman, and Michael Weinberg, "Know Your Limits: Considering the Role of Data Caps and Usage Based Billing in Internet Access Service," *Public Knowledge,* April 23, 2012, http://www.publicknowledge.org/documents/know-your-limits-considering-the-role-of-data-caps-and-usage-based-billing.

14. The Broadband Commission, *The State of Broadband 2014: Broadband for All* (Geneva, Switzerland: ITU and UNESCO, 2014), http://www.broadbandcommission.org/Documents/reports/bb-annualreport2014.pdf.

15. International Telecommunication Union, "The World in 2013: ICT Facts and Figures" (Geneva, 2013), http://www.itu.int/en/ITU-D/Statistics/Documents/facts/ICTFactsFigures2013-e.pdf.

16. GSMA and A. T. Kearney, "The Mobile Economy 2013" (London, 2013), www.atkearney.com/documents/10192/760890/The_Mobile_Economy_2013.pdf.

17. Mary Meeker, "Internet Trends 2014: Code Conference," *KPCB News,* May 28, 2014, p. 72, http://kpcbweb2.s3.amazonaws.com/files/85/Internet_Trends_2014_vFINAL_-_05_28_14-_PDF.pdf?1401286773.

18. Aileen Agüero, Harsha de Silva, and Juhee Kang, "Bottom of the Pyramid Expenditure Patterns on Mobile Services in Selected Emerging Asian Countries," *Information Technologies & International Development* 7, no. 3 (2011): 19–32.

19. Nithya Sambasivan, Paul Lee, Greg Hecht, Paul M. Aoki, Maria-Ines Carrera, Jenny Chen, David Pablo Cohn, Pete Kruskall, Everett Wetchler, Michael Youssefmir, and Astrid Twenebowa Larssen, "Chale, How Much It Cost to Browse?: Results from a Mobile Data Price Transparency Trial in Ghana," in *Proceedings of the Sixth International Conference on Information and Communication Technologies and Development Full Papers—ICTD '13,* vol. 1 (New York: ACM Press, 2013), 13–23, doi:10.1145/2516604.2516607; Rade Stanojevic, Vijay Erramilli, and Konstantina Papagiannaki, "Cognitive Bias in Network Services," in *Proceedings of the 11th ACM Workshop on Hot Topics in Networks—HotNets-XI* (New York: ACM Press, 2012), 49–54, doi:10.1145/2390231.2390240.

20. OAfrica, "Comparing African Pre-Paid Mobile Broadband Plans," *OAfrica Blog,* September 24, 2012, http://www.oafrica.com/mobile/african-pre-paid-mobile-broadband-plans/.

21. Joss Gillet, Tim Hatt, and Barbara Arese Lucini, "Tailoring Mobile Internet Tariffs for Prepaid Users—a Balancing Act," *GSMA Intelligence Reports* (London, December 2013), https://gsmaintelligence.com/files/analysis/?file=131205-prepaid-data-tariffs.pdf; Kevin P. Donovan and Jonathan Donner, "A Note on the

Availability (and Importance) of Pre-Paid Mobile Data in Africa," in *Proceedings of the 2nd International Conference on M4D: Mobile Communication Technology for Development*, ed. Jakob Svensson and Gudrun Wicander (Karlstad, Sweden: Karlstad University, 2010), 263–267.

22. International Telecommunication Union, *Measuring the Information Society Report 2014* (Geneva: ITU, 2014), p. 129, http://www.itu.int/en/ITU-D/Statistics/Documents/publications/mis2014/MIS2014_without_Annex_4.pdf.

23. Kartikay Mehrotra, "Telecom Tariff War Shifts to Data as Smartphones Surge," *LiveMint*, July 15, 2013, http://www.livemint.com/Consumer/1nrqfjE9qecR2LJYz6UqYO/Smartphone-surge-spurs-data-war-as-tariffs-slashed.html.

24. Chetty et al., "'You're Capped!'"

25. Gary W. Pritchard and John Vines, "Digital Apartheid: An Ethnographic Account of Racialised HCI in Cape Town Hip-Hop," in *Proceedings of the SIGCHI Conference on Human Factors in Computing Systems—CHI '13* (New York: ACM Press, 2013), 2537–2346, doi:10.1145/2470654.2481350.

26. Ericsson, "Ericsson Mobility Report: On the Pulse of the Networked Society" (Stockholm, November 2014), http://www.ericsson.com/res/docs/2014/ericsson-mobility-report-november-2014.pdf, 23.

27. Stanojevic, Erramilli, and Papagiannaki, "Cognitive Bias in Network Services."

28. Anne Oeldorf-Hirsch, Jonathan Donner, and Edward Cutrell, "How Bad Is Good Enough?," in *Proceedings of the 7th Nordic Conference on Human–Computer Interaction Making Sense Through Design—NordiCHI '12* (New York: ACM Press, 2012), 49–58, doi:10.1145/2399016.2399025.

29. Chetty et al., "'You're Capped!'"

30. Vanessa Daly, "Wi-Fi Grows to over 50% of Mobile Web Connections," *Bango.com Blog*, February 2, 2011, http://news.bango.com/2011/02/02/wi-fi-grows-to-over-50-percent/; Mobidia, "Managed Wi-Fi Hotspot Usage," *Understanding Mobile Data Blog*, February 21, 2013, http://mobidia.blogspot.com/2013/02/managed-wi-fi-hotspot-usage-over-past.html; Cisco, "Cisco Visual Networking Index: Global Mobile Data Traffic Forecast Update, 2013–2018" (San Jose CA, 2014), https://web.archive.org/web/20140302213838/http://www.cisco.com/c/en/us/solutions/collateral/service-provider/visual-networking-index-vni/white_paper_c11-520862.html?

31. Katy E. Pearce, "Convergence through Mobile Peer-to-Peer File Sharing in the Republic of Armenia," *International Journal of Communication* 5 (2011): 511–528; Thomas N. Smyth, Satish Kumar, Indrani Medhi, and Kentaro Toyama, "Where There's a Will There's a Way: Mobile Media Sharing in Urban India," in *Proceedings of the 28th International Conference on Human Factors in Computing Systems—CHI '10*

(New York: ACM Press, 2010), 753–762, doi:10.1145/1753326.1753436; Marion Walton, Gary Marsden, Silke Hassreiter, and Sena Allen, "Degrees of Sharing: Proximate Media Sharing and Messaging by Young People in Khayelitsha," in *Proceedings of the 14th International Conference on Human–Computer Interaction with Mobile Devices and Services—MobileHCI '12* (New York: ACM Press, 2012), 403–412, doi:10.1145/2371574.2371636.

32. GSMA Mobile for Development Intelligence, *Scaling Mobile for Development: Harness the Opportunity* (London, August 2013), https://gsmaintelligence.com/research/?file=130828-scaling-mobile.pdf.

33. Sambasivan et al., "Chale, How Much It Cost to Browse?"

34. Taylor, "Pre-Paid Literacy." See also Duncan, "Mobile Network Society?"

35. Paul Thurrott, "Windows Phone 8: Data Sense," *Paul Thurrott's Supersite for Windows*, December 10, 2012, http://winsupersite.com/windows-phone/windows-phone-8-data-sense.

36. Sambasivan et al., "Chale, How Much It Cost to Browse?"

37. Sindhura Chava, Rachid Ennaji, Jay Chen, and Lakshminarayanan Subramanian, "Cost-Aware Mobile Web Browsing," *IEEE Pervasive Computing* 11, no. 3 (March 1, 2012): 34–42, doi:10.1109/MPRV.2012.19.

38. Chetty et al., "'You're Capped!'"

39. Sambasivan et al., "Chale, How Much It Cost to Browse?"

40. Ibid.

41. Donovan and Donner, "A Note on the Availability (and Importance) of Pre-Paid Mobile Data in Africa"; Steve Costello, "Ovum: Emerging Market Operators Must 'Embrace' Prepaid Base & Simplify Tariffs," *Mobile World Live*, August 20, 2013, http://www.mobileworldlive.com/ovum-emerging-market-operators-must-embrace-prepaid-base-simplify-tariffs; Gillet, Hatt, and Lucini, "Tailoring Mobile Internet Tariffs for Prepaid Users."

42. Susan P. Wyche, Thomas N. Smyth, Marshini Chetty, Paul M. Aoki, and Rebecca E. Grinter, "Deliberate Interactions," in *Proceedings of the 28th International Conference on Human Factors in Computing Systems—CHI '10* (New York: ACM Press, 2010), 2593–2602, doi:10.1145/1753326.1753719.

43. Susan P. Wyche and Laura L. Murphy, "Powering the Cellphone Revolution: Findings from Mobile Phone Charging Trials in off-Grid Kenya," in *Proceedings of the SIGCHI Conference on Human Factors in Computing Systems—CHI '13* (New York: ACM Press, 2013), 1959–1968, doi:10.1145/2470654.2466260.

44. Jonathan Donner, Shikoh Gitau, and Gary Marsden, "Exploring Mobile-Only Internet Use: Results of a Training Study in Urban South Africa," *International Journal of Communication* 5 (2011): 574–597.

45. Marion Walton, "Pavement Internet: Mobile Media Economies and Ecologies in South Africa," in *Routledge Companion to Mobile Media*, ed. Gerard Goggin and Larissa Hjorth (London: Routledge, 2014), 450–461; p. 451.

46. Richard Harper, Tim Regan, Shahram Izadi, Kharsim Al Mosawi, Mark Rouncefield, and Simon Rubens, "Trafficking: Design for the Viral Exchange of TV Content on Mobile Phones," in *Proceedings of the 9th International Conference on Human Computer Interaction with Mobile Devices and Services—MobileHCI '07*, vol. 7 (New York: ACM Press, 2007), 249–256, doi:10.1145/1377999.1378015; Neha Kumar, Gopal Chouhan, and Tapan Parikh, "Folk Music Goes Digital in India," in *Proceedings of the 2011 Annual Conference on Human Factors in Computing Systems—CHI '11* (New York: ACM Press, 2011), 1423–1432, doi:10.1145/1978942.1979151; Jay Chen, Lakshminarayanan Subramanian, and Kentaro Toyama, "Web Search and Browsing Behavior under Poor Connectivity," in *Proceedings of the 27th International Conference Extended Abstracts on Human Factors in Computing Systems—CHI EA '09* (New York: ACM Press, 2009), 3473–3478, doi:10.1145/1520340.1520505.

47. Roberto Baldwin, "Facebook Updates iOS and Android Apps with Nearby Friends Invites and Offline Likes," *Thenextweb.com*, 2014, http://thenextweb.com/apps/2014/06/12/facebook-updates-ios-android-apps-nearby-friends-invitations-offline-posts-respectively/.

48. Gary Marsden, Edward Cutrell, Matt Jones, Amit A. Nanavati, and Nitendra Rajput, "Making Technology Invisible in the Developing World," *Computer* 45, no. 4 (April 2012): 82–85, doi:10.1109/MC.2012.141.

49. Prasanta Bhattacharya and William Thies, "Computer Viruses in Urban Indian Telecenters: Characterizing an Unsolved Problem," in *Proc ACM Workshop on Networked Systems for Developing Regions (NSDR 2011)* (New York: ACM, 2011), 1–6.

50. Julia Bello-Bravo and Ibrahim Baoua, "Animated Videos as a Learning Tool in Developing Nations: A Pilot Study of Three Animations in Maradi and Surrounding Areas in Nigee," *Electronic Journal of Information Systems in Developing Countries* 55 (2012): 1–12; Andrew Maunder, Gary Marsden, and Richard Harper, "Creating and Sharing Multi-Media Packages Using Large Situated Public Displays and Mobile Phones," in *Proceedings of the 9th International Conference on Human Computer Interaction with Mobile Devices and Services*, 2007, 222–25, doi:10.1145/1377999.1378010.

51. ITWeb, "MTN Intros Cheaper Prepaid Data Bundles," *ITWeb*, 2014, http://www.itweb.co.za/?id=138550:MTN-intros-cheaper-prepaid-data-bundles.

52. Soumya Sen, Carlee Joe-Wong, Sangtae Ha, and Mung Chiang, "A Survey of Smart Data Pricing," *ACM Computing Surveys* 46, no. 2 (November 1, 2013): 1–37, doi:10.1145/2543581.2543582.

53. Rich Edmonds, "Nokia Works with Indian Mobile Operators to Provide Free Data for Lumia Owners," *Windows Phone Central*, August 26, 2013, http://www.wpcentral.com/nokia-works-indian-mobile-operators-provide-free-data-lumia-owners.

54. Nathan Eagle, "Txteagle: Mobile Crowdsourcing," in *IDGD '09 Proceedings of the 3rd International Conference on Internationalization, Design and Global Development: Held as Part of HCI International 2009* (Heidelberg: Springer-Verlag Berlin, 2009), 447–456, doi:10.1007/978-3-642-02767-3_50.

55. Cyrus Farivar, "Belgium-Based, Viking-Named Startup Wants to Vanquish Mobile Data Costs," *Ars Technica*, February 4, 2013, http://arstechnica.com/business/2013/02/belgian-based-viking-named-startup-wants-to-vanquish-mobile-data-costs/-p3n.

56. Nithya Sambasivan and Edward Cutrell, "Understanding Negotiation in Airtime Sharing in Low-Income Microenterprises," in *Proceedings of the 2012 ACM Annual Conference on Human Factors in Computing Systems—CHI '12* (New York: ACM Press, 2012), 791–800, doi:10.1145/2207676.2207791.

57. Martin Cave and Windfred Mfuh, "Rethinking Mobile Regulation for the Data Age," in Kirk, Bratt, and Coyle, *Making Broadband Accessible for All* (London: Vodafone Group, 2011), 41–48.

58. Anton Troianovski, "AT&T May Try Billing App Makers," *The Wall Street Journal*, February 28, 2012, http://online.wsj.com/article/SB10001424052970204653604577249080966030276.html.

59. Kristofer Kimbler and Mac Taylor, "Value Added Mobile Broadband Services Innovation Driven Transformation of the 'Smart Pipe,'" in *2012 16th International Conference on Intelligence in Next Generation Networks* (New York: IEEE, 2012), 30–34, doi:10.1109/ICIN.2012.6376030.

60. Sid Murlidhar, "Fast and Free Facebook Mobile Access with 0.facebook.com," *The Facebook Blog*, May 18, 2010, http://blog.facebook.com/blog.php?post=391295167130.

61. Max Fisher, "Facebook's Amazing Growth in the Developing World," *The Atlantic*, May 18, 2012, http://www.theatlantic.com/international/archive/2012/05/facebooks-amazing-growth-in-the-developing-world/257392/; Christopher Mims, "Facebook's Plan to Find Its Next Billion Users: Convince Them the Internet and Facebook Are the Same," *Quartz*, September 24, 2012, http://qz.com/5180/facebooks-plan-to-find-its-next-billion-users-convince-them-the-internet-and-facebook-are-the-same/.

62. Emil Protalinski, "Facebook Launches Facebook for Every Phone App," *ZDNet*, July 12, 2012, http://www.zdnet.com/blog/facebook/facebook-launches-facebook -for-every-phone-app/2166.

63. Daniel Cooper, "Facebook and 18 Carriers to Offer Discounted Mobile Messaging Data in 14 Countries," *Engadget*, February 13, 2013, http://www.engadget.com/ 2013/02/25/facebook-carrier-deal/.

64. CIOL Bureau, "With Airtel and Google's Free Zone, Access Web Pages Free," *CIOL*, June 26, 2013, http://www.ciol.com/ciol/news/190657/with-airtel-googles -free-zone-access-web-pages-free.

65. Sarah Perez, "Twitter's 'Zero' Service Lets Emerging Markets Tweet For Free," *Techcrunch*, 2014, http://techcrunch.com/2014/05/29/twitters-emerging-market -strategy-includes-its-own-version-of-a-facebook-zero-like-service-called-twitter- access/.

66. Kal Wadhwa, "Getting Wikipedia to the People Who Need It Most," *Wikimedia Foundation Blog*, February 22, 2013, http://blog.wikimedia.org/2013/02/22/getting -wikipedia-to-the-people-who-need-it-most/.

67. Joshua Goodman and Barbara Ortutay, "Mark Zuckerberg Visits Colombia, Launches Free Internet Project," *San Jose Mercury News*, January 15, 2015, http:// www.mercurynews.com/business/ci_27321291/mark-zuckerberg-visits-colombia -launches-free-internet-project.

68. Issie Lapowsky, "Zuckerberg Expands Internet.org After Net Neutrality Uproar," *Wired*, May 4, 2015, http://www.wired.com/2015/05/internet-org-expands-net -neutrality/.

69. Jon Brodkin, "AT&T Has 10 Businesses Paying for Data Cap Exemptions, and Wants More," *ArsTechnica*, January 6, 2015, http://arstechnica.com/ business/2015/01/att-has-10-businesses-paying-for-data-cap-exemptions-and-wants -more/.

70. Marion Walton, "Mobile Literacies and South African Teens: Leisure Reading, Writing, and MXit Chatting for Teens in Langa and Guguletu," *Report for the Shuttleworth Foundation* (Cape Town, December 2009), https://m4lit.files.wordpress.com/ 2010/03/m4lit_mobile_literacies_mwalton_20101.pdf.

Chapter 10

1. Jonathan Donner and Marion Walton, "Your Phone Has Internet—Why Are You at a Library PC? Re-Imagining Public Access for the Mobile Internet Era," in *Proceedings of INTERACT 2013: 14th IFIP TC 13 International Conference* (Berlin: Springer, 2013), 347–364, doi:10.1007/978-3-642-40483-2_25.

2. Jeff Grubb, "Minecraft and Candy Crush Saga Dominate Apple's iOS Charts for 2013," *VentureBeat*, December 17, 2013, http://venturebeat.com/2013/12/17/minecraft-and-candy-crush-saga-dominate-apples-ios-charts-for-2013/.

3. Simon Khalaf, "Flurry Five-Year Report: It's an App World. The Web Just Lives in It," *Flurry Blog*, April 3, 2013, http://flurrymobile.tumblr.com/post/115188952445/flurry-five-year-report-its-an-app-world-the.

4. Don Slater, *New Media, Development & Globalization* (Cambridge, UK: Polity Press, 2013), 99–129.

5. Susan P. Wyche, Andrea Forte, and Sarita Yardi Schoenebeck, "Hustling Online: Understanding Consolidated Facebook Use in an Informal Settlement in Nairobi," in *Proceedings of the SIGCHI Conference on Human Factors in Computing Systems—CHI '13* (New York: ACM Press, 2013), 2823–2832, doi:10.1145/2470654.2481391.

6. Mary Meeker, "Internet Trends 2014: Code Conference," *KPCB News*, May 28, 2014, p. 62, http://kpcbweb2.s3.amazonaws.com/files/85/Internet_Trends_2014_vFINAL_-_05_28_14-_PDF.pdf?1401286773.

7. Yochai Benkler, *The Wealth of Networks: How Social Production Transforms Markets and Freedom* (New Haven, CT: Yale University Press, 2006).

8. Jason Wilson, "3G to Web 2.0? Can Mobile Telephony Become an Architecture of Participation?," *Convergence: The International Journal of Research into New Media Technologies* 12, no. 2 (May 1, 2006): 229–242, doi:10.1177/1354856506066122.

9. Axel Bruns, *Blogs, Wikipedia, Second Life, and Beyond: From Production to Produsage* (New York: Peter Lang, 2008).

10. Niklas Woermann, "On the Slope Is on the Screen: Prosumption, Social Media Practices, and Scopic Systems in the Freeskiing Subculture," *American Behavioral Scientist* 56, no. 4 (March 21, 2012): 618–640, doi:10.1177/0002764211429363.

11. Payal Arora, "The Leisure Divide: Can the 'Third World' Come Out to Play?," *Information Development* 28, no. 2 (February 2, 2012): 93–101, doi:10.1177/0266666911433607; Beth E. Kolko and Cynthia Putnam, "Computer Games in the Developing World: The Value of Non-Instrumental Engagement with ICTs, or Taking Play Seriously," *2009 International Conference on Information and Communication Technologies and Development ICTD*, April 2009, 46–55, doi:10.1109/ICTD.2009.5426705.

12. Alex S. Taylor and Richard Harper, "The Gift of the Gab?: A Design Oriented Sociology of Young People's Use of Mobiles," *Computer Supported Cooperative Work (CSCW)* 12, no. 3 (September 2003): 267–296, doi:10.1023/A:1025091532662.

13. Pedro Ferreira and Kristina Höök, "Appreciating Plei-Plei around Mobiles: Playfulness in Rah Island," in *Conference on Human Factors in Computing Systems—Proceedings* (Austin, TX: ACM, 2012), 2015–2024, doi:10.1145/2207676.2208348.

14. James E. Katz and Mark Aakhus, "Conclusion: Making Meaning of Mobiles: A Theory of Apparatgeist," in *Perpetual Contact: Mobile Communication, Private Talk, and Public Performance*, ed. James E. Katz and Mark Aakhus (Cambridge, UK: Cambridge University Press, 2002), 301–318.

15. Dorothea Kleine, *Technologies of Choice? ICTs, Development, and the Capabilities Approach* (Cambridge, MA: MIT Press, 2013), p. 9.

16. BizTechAfrica, "SA Author Writes Acclaimed Novel Using BlackBerry Smartphone," *BizTechAfrica*, October 15, 2014, http://www.biztechafrica.com/article/sa-author-writes-acclaimed-novel-using-blackberry-/8961/.

17. Yonnie Kim, "Genealogy of Mobile Creativity," in *Routledge Companion to Mobile Media*, ed. Gerard Goggin and Larissa Hjorth (New York: Routledge, 2014), 216–224.

18. Hazal Kirci, "The Tales Teens Tell: What Wattpad Did for Girls," *The Guardian*, August 16, 2014, http://www.theguardian.com/technology/2014/aug/16/teen-writing-reading-wattpad-young-adults.

19. Max Schleser, "A Decade of Mobile Moving-Image Practice," in *Routledge Companion to Mobile Media*, ed. Gerard Goggin and Larissa Hjorth (New York: Routledge, 2014), 157–170.

20. Darrell Etherington, "Apple Gets Serious about the iPad's Creative Power in New Ad," *Techcrunch*, January 12, 2014, http://techcrunch.com/2014/01/12/apple-your-verse-ipad-ad/.

21. Paul Thurrott, "Microsoft Launches Office Apps for Android Tablets and iPhone, Updates Office for iPad," *Paul Thurrott's Supersite for Windows* (blog), November 2014, http://winsupersite.com/office-365/microsoft-launches-office-apps-android-tablets-and-iphone-updates-office-ipad.

22. Matthew Panzarino, "The Tablet Is the New General Purpose Computer," *Techcrunch*, January 18, 2014, http://techcrunch.com/2014/01/18/the-tablet-is-the-new-general-purpose-computer/; Josh Constine, "Quip Is a Beautiful New Mobile-First Word Processor from Ex-Facebook CTO Bret Taylor," *Techcrunch*, July 30, 2013, http://techcrunch.com/2013/07/30/quip-mobile-word-processor/.

23. Paul D. Miller and Svitlana Matviyenko, "Introduction," in *The Imaginary App*, ed. Paul D. Miller and Svitlana Matviyenko (Cambridge, MA: MIT Press, 2014), xi.

24. Jen Schradie, "The Digital Production Gap: The Digital Divide and Web 2.0 Collide," *Poetics* 39, no. 2 (April 2011): 145–168, doi:10.1016/j.poetic.2011.02.003, p. 160; Christian Fuchs, "Theorising and Analysing Digital Labour: From Global Value Chains to Modes of Production," *The Political Economy of Communication* 1, no. 2 (2013), np, http://www.polecom.org/index.php/polecom/article/view/19/195.

25. Richard A. Duncombe, "Using the Livelihoods Framework to Analyze ICT Applications for Poverty Reduction through Microenterprise," *Information Technologies &*

International Development 3, no. 3 (March 2007): 81–100, doi:10.1162/itid.2007 .3.3.81.

26. Jonathan Donner and Marcela X. Escobari, "A Review of Evidence on Mobile Use by Micro and Small Enterprises in Developing Countries," *Journal of International Development* 22, no. 5 (2010): 641–658, doi:10.1002/jid.1717; Richard A. Duncombe, "Understanding the Impact of Mobile Phones on Livelihoods in Developing Countries," *Development Policy Review* 32, no. 5 (2014): 567–588, doi:10.1111/dpr.12073.

27. Katy E. Pearce and Ronald E. Rice, "Digital Divides from Access to Activities: Comparing Mobile and Personal Computer Internet Users," *Journal of Communication* 63, no. 4 (2013): 721–744, doi:10.1111/jcom.12045; Katy E. Pearce, Janine Slaker, and Nida Ahmad, "Is Your Web Everyone's Web? Theorizing the Web through the Lens of the Device Divide," *Theorizing the Web 2012*, April 12, 2012, http://fr.slideshare.net/katyp1/katy-pearce-ttw12.

28. Alexander van Deursen and Jan van Dijk, "The Digital Divide Shifts to Differences in Usage," *New Media & Society* 16, no. 3 (June 7, 2013): 507–526, doi:10.1177/1461444813487959.

29. Nicole Zillien and Eszter Hargittai, "Digital Distinction: Status-Specific Types of Internet Usage," *Social Science Quarterly* 90, no. 2 (June 2009): 274–291, doi:10.1111/j.1540-6237.2009.00617.x

30. Pearce and Rice, "Digital Divides from Access to Activities"; Pearce, Slaker, and Ahmad, "Is Your Web Everyone's Web?"

31. Jonathan Donner and Andrew Maunder, "Beyond the Phone Number: Challenges of Representing Informal Microenterprise on the Internet," in *Living Inside Mobile Social Information*, ed. James E. Katz (Dayton, Ohio: Greyden Press, 2014), 159–192.

32. Sherry R. Arnstein, "A Ladder Of Citizen Participation," *Journal of the American Institute of Planners* 35, no. 4 (July 1969): 216, doi:10.1080/01944366908977225.

33. Mark Thompson, "ICT and Development Studies: Towards Development 2.0," *Journal of International Development* 20, no. 6 (August 2008): 821–835, doi:10.1002/ jid.1498.

34. Richard Heeks, "ICT4D 2.0: The Next Phase of Applying ICT for International Development," *Computer* 41, no. 6 (June 2008): 26–33, doi:10.1109/MC.2008.192.

35. Aaron Sankin, "Mobile-Only Internet Users Face a Harsh New Digital Divide," *The Daily Dot*, November 22, 2013, http://www.dailydot.com/lifestyle/digital-divide -mobile-internet-challenges/; Schradie, "The Digital Production Gap."

36. Schradie, "The Digital Production Gap."

37. van Deursen and van Dijk, "The Digital Divide Shifts to Differences in Usage."

38. Darrell Etherington, "Forrester Finds That Despite the Tablet Invasion of the Workplace, Workers Would Prefer a Keyboard, too," *Techcrunch*, August 16, 2013, http://techcrunch.com/2013/08/16/forrester-finds-that-despite-the-tablet-invasion -of-the-workplace-workers-would-prefer-a-keyboard-too.

39. Shikoh Gitau, Gary Marsden, and Jonathan Donner, "After Access: Challenges Facing Mobile-Only Internet Users in the Developing World," in *Proceedings of the 28th International Conference on Human Factors in Computing Systems—CHI '10* (New York: ACM, 2010), 2603–2606, doi:10.1145/1753326.1753720.

40. Marion Walton and Jonathan Donner, "Read-Write-Erase: Mobile-Mediated Publics in South Africa's 2009 Elections," in *Mobile Communication: Dimensions of Social Policy*, ed. James E. Katz (New Brunswick, NJ: Transaction Publishers, 2011), 117–132.

41. danah m. boyd and Nicole B. Ellison, "Social Network Sites: Definition, History, and Scholarship," *Journal of Computer-Mediated Communication* 13, no. 1 (October 17, 2007): 210–230, doi:10.1111/j.1083-6101.2007.00393.x.

42. Marion Walton and Pierrinne Leukes, "Prepaid Social Media and Mobile Discourse in South Africa," *Journal of African Media Studies* 5, no. 2 (2013): 149–167, doi:10.1386/jams.5.2.149_1, p. 159.

43. Lincoln Dahlberg, "The Internet and Democratic Discourse: Exploring The Prospects of Online Deliberative Forums Extending the Public Sphere," *Information, Communication & Society* 4, no. 4 (2001): 615–633, doi:10.1080/13691180110097030; Zizi Papacharissi, "The Virtual Sphere: The Internet as a Public Sphere," *New Media & Society* 4, no. 1 (2002): 9–27, doi:10.1177/14614440222226244.

44. Schradie, "The Digital Production Gap."

45. Microsoft Research, "Microsoft Brings World's Fastest Texting to Windows Phone 8.1," *Microsoft Research Blog*, April 4, 2014, http://research.microsoft.com/ en-us/news/features/wordflow-040414.aspx.

46. Leslie L. Dodson, S. Revi Sterling, and John K. Bennett, "Minding the Gaps," in *Proceedings of the Sixth International Conference on Information and Communication Technologies and Development Full Papers—ICTD '13*, vol. 1 (New York: ACM Press, 2013), 79–88, doi:10.1145/2516604.2516626.

47. Jan Chipchase, "Understanding Non-Literacy as a Barrier to Mobile Phone Communication" (Nokia Research Center, 2005); Indrani Medhi, Somani Patnaik, Emma Brunskill, S. N. Nagasena Gautama, William Thies, and Kentaro Toyama, "Designing Mobile Interfaces for Novice and Low-Literacy Users," *ACM Transactions on Computer-Human Interaction* 18, no. 1 (2011): 1–28, doi:10.1145/1959022.1959024; Indrani Medhi, Aman Sagar, and Kentaro Toyama, "Text-Free User Interfaces for Illiterate and Semiliterate Users," *Information Technologies & International Development* 4, no. 1 (2007): 37–50, doi:10.1162/itid.2007.4.1.37.

48. Aditya Vashistha and William Thies, "How Should Users Convey Their Location to an Interactive Voice Response System?," in *Proceedings of the 4th Annual Symposium on Computing for Development—ACM DEV-4 '13* (New York: ACM Press, 2013), doi:10.1145/2537052.2537078.

49. Nicola J. Bidwell and Masbulele Jay Siya, "Situating Asynchronous Voice in Rural Africa," in *Human–Computer Interaction—INTERACT 2013: 14th IFIP TC 13 International Conference, Cape Town, South Africa, September 2–6, 2013, Proceedings, Part III* (Berlin, Heidelberg: Springer, 2013), 36–53, doi:10.1007/978-3-642-40477 -1_3.

50. Haohan Wang, Agha Ali Raza, Yibin Lin, and Roni Rosenfeld, "Behavior Analysis of Low-Literate Users of a Viral Speech-Based Telephone Service," in *Proceedings of the 4th Annual Symposium on Computing for Development—ACM DEV-4 '13* (New York: ACM Press, 2013), 1–9, doi:10.1145/2537052.2537062.

51. Neil Patel, Deepti Chittamuru, Anupam Jain, Paresh Dave, and Tapan S. Parikh, "Avaaj Otalo: A Field Study of an Interactive Voice Forum for Small Farmers in Rural India," in *Proceedings of the 28th International Conference on Human Factors in Computing Systems—CHI '10* (New York: ACM Press, 2010), 733–742, doi:10.1145/ 1753326.1753434.

52. Jahanzeb Sherwani, Nosheen Ali, Sarwat Mirza, Anjum Fatma, Yousuf Memon, Mehtab Karim, Rahul Tongia, and Roni Rosenfeld, "Healthline: Speech-Based Access to Health Information by Low-Literate Users," in *Proceedings of the International Conference on Information and Communication Technologies and Development—ICTD 2007* (New York: IEEE Press, 2007), 131–139, doi:10.1109/ICTD.2007.4937399.

53. M. Metcalfe, "Development and Oral Technologies," *Information Technology for Development* 13, no. 2 (2007): 199. Preeti Mudliar, Jonathan Donner, and William Thies, "Emergent Practices around CGNet Swara: A Voice Forum for Citizen Journalism in Rural India," *Information Technologies & International Development* 9, no. 2 (2013): 65–79.

54. Sheetal K. Agarwal, Arun Kumar, Amit Ani Nanavati, and Nitendra Rajput, "Content Creation and Dissemination by-and-for Users in Rural Areas," in *Proceedings of the 2009 International Conference on Information and Communication Technologies and Development—ICTD 2009* (New York: IEEE Press, 2009), 56–65, doi:10.1109/ ICTD.2009.5426702; Arun Kumar, Sheetal K. Agarwal, and Priyanka Manwani, "The Spoken Web Application Framework," in *Proceedings of the 2010 International Cross Disciplinary Conference on Web Accessibility (W4A)—W4A '10* (New York: ACM Press, 2010), np, doi:10.1145/1805986.1805990.

55. Roberto Pieraccini, *The Voice in the Machine: Building Computers That Understand Speech* (Cambridge, MA: MIT Press, 2012).

56. Apple, "About Siri," *Apple Support Pages*, October 13, 2014, https://support .apple.com/en-us/HT204389; Amos Cruz, "How Can We Make Cloud Solutions Relevant in the Offline World?," *ICTWorks* (blog), October 17, 2012, http://www .ictworks.org/2012/10/17/how-can-we-make-cloud-solutions-relevant-offline -world/.

57. Sarah Kendzior, "Lost in Google's Translation," *Registan*, April 4, 2012, http:// registan.net/2012/04/30/lost-in-googles-translation/.

58. Brent Hecht and Darren Gergle, "The Tower of Babel Meets Web 2.0: User-Generated Content and Its Applications in a Multilingual Context," in *Proceedings of the 28th International Conference on Human Factors in Computing Systems—CHI '10* (New York: ACM Press, 2010), 291–300, doi:10.1145/1753326.1753370; Mark Surman, Corina Gardner, and David Ascher, "Local Content, Smartphones, and Digital Inclusion," *Innovations: Technology, Governance, Globalization* 9, no. 3–4 (July 2014): 68, doi:10.1162/inov_a_00217.

59. Madeline Plauché and Udhyakumar Nallasamy, "Speech Interfaces for Equitable Access to Information Technology," *Information Technologies & International Development* 4, no. 1 (2007): 69–86, doi:10.1162/itid.2007.4.1.69; Iris Orriss, "The Internet's Language Barrier," *Innovations: Technology, Governance, Globalization* 9, no. 3–4 (July 2014): 123–126, doi:10.1162/inov_a_00223.

60. Amos Cruz, "How Can We Make Cloud Solutions Relevant in the Offline World?," *ICTWorks* (blog), October 17, 2012, http://www.ictworks.org/2012/10/17/ how-can-we-make-cloud-solutions-relevant-offline-world/.

61. Julie Coiro, Michele Knobel, Colin Lankshear, and Donald J. Leu, eds. *Handbook of Research on New Literacies* (New York: Routledge, 2008); Sonia Livingstone, "Media Literacy and the Challenge of New Information and Communication Technologies," *The Communication Review* 7, no. 1 (January 2004): 3–14, doi:10.1080/ 10714420490280152; Eszter Hargittai and Yuli Patrick Hsieh, "Digital Inequality," in *The Oxford Handbook of Internet Studies*, ed. William H. Dutton (Oxford, UK: Oxford University Press, 2013), 129–150.

62. Annita Fjuk, Anniken Furberg, Hanne Cecilie Geirbo, and Per Helmersen, "New Artifacts—New Practices: Putting Mobile Literacies into Focus," *Nordic Journal of Digital Literacy* 3 (2008): 21–38; Andrea Kavanaugh, Anita Puckett, and Deborah Tatar, "Scaffolding Technology for Low Literacy Groups: From Mobile Phone to Desktop PC?," *International Journal of Human–Computer Interaction* 29, no. 4 (March 2013): 274–288, doi:10.1080/10447318.2013.765766.

63. K. V. Lexander, "Texting and African Language Literacy," *New Media & Society* 13, no. 3 (2011): 427–442, doi:10.1177/1461444810393905; Fie Velghe, "Literacy Acquisition, Informal Learning and Mobile Phones in a South African Township," in *Proceedings of the Sixth International Conference on Information and Communication*

Technologies and Development Full Papers—ICTD '13, vol. 1 (New York: ACM Press, 2013), 89–99, doi:10.1145/2516604.2516615.

64. Anja Venter, "Creative Participation and Mobile Ecologies among Resource-Constrained Aspirant Designers in Cape Town, South Africa," in *Proceedings of the 13th Participatory Design Conference: Short Papers, Industry Cases, Workshop Descriptions, Doctoral Consortium Papers, and Keynote Abstracts—PDC '14*, vol. 2 (New York: ACM, 2014), 225–228, doi:10.1145/2662155.2662241.

65. Calvin Taylor, "Pre-Paid Literacy: Negotiating the Cost of Adolescent Mobile Technology Use," *English in Australia* 44, no. 2 (2009): 26–34.

66. Kristen Yarmey, "Student Information Literacy in the Mobile Environment," *EDUCAUSE Quarterly* 34, no. 1 (2011), np, www.educause.edu/ero/article/student -information-literacy-mobile-environment.

67. Gitau, Marsden, and Donner, "After Access."

68. Marion Walton, "Mobile Literacies: Messaging, Txt and Social Media in m4Lit," in *Multimodal Approaches to Research and Pedagogy: Recognition, Resources and Access*, ed. A. Archer and D. Newfield (London: Routledge, 2014), 108–127.

69. Jerry Watkins, Kathi R. Kitner, and Dina Mehta, "Mobile and Smartphone Use in Urban and Rural India," *Continuum* 26, no. 5 (October 2012): 685–697, doi: 10.1080/10304312.2012.706458.

70. Mirca Madianou and Daniel Miller, "Polymedia: Towards a New Theory of Digital Media in Interpersonal Communication," *International Journal of Cultural Studies* 16, no. 2 (August 22, 2012): 169–187, doi:10.1177/1367877912452486.

71. Henry Jenkins, *Confronting the Challenges of Participatory Culture-Media Education for the 21st Century* (Cambridge, MA: MIT Press, 2009).

72. Emrys Schoemaker, "The Mobile Web: Amplifying, but Not Creating, Change-makers," *Innovations: Technology, Governance, Globalization* 9, no. 3–4 (July 2014): 75–85, doi:10.1162/inov_a_00218; Jonathan Donner and Patricia Mechael, eds., *mHealth in Practice: Mobile Technology for Health Promotion in the Developing World* (London: Bloomsbury Academic, 2012).

73. Mozilla, "Web Literacy Map—1.1.0," *Mozilla Webmaker*, May 10, 2015, https://webmaker.org/en-US/literacy.

74. Jon Brodkin, "Ubuntu Smartphones Will Cost $200–$400," *Ars Technica*, March 12, 2014, http://arstechnica.com/gadgets/2014/03/ubuntu-smartphones-will -cost-200-400/.

75. James Rogerson, "Windows Phone 10 Release Date, News and Features," *Techradar*, January 21, 2015, http://www.techradar.com/news/software/operating-systems/windows-phone-9-nine-things-we-want-to-see-1092322.

76. Preeti Mudliar and Joyojeet Pal, "ICTD in the Popular Press," in *Proceedings of the Sixth International Conference on Information and Communication Technologies and Development Full Papers—ICTD '13*, vol. 1 (New York: ACM Press, 2013), 43–54, doi:10.1145/2516604.2516629; Christopher Peri, "Hands-on with the $35 Aakash2 Tablet: I Want One," *VentureBeat*, September 18, 2012.

77. Ron Amadeo, "Samsung's Galaxy Tab3 Lite Should Be Its Cheapest Tablet Ever," *Ars Technica*, January 16, 2014, http://arstechnica.com/gadgets/2014/01/samsungs-galaxy-tab3-lite-should-be-its-cheapest-tablet-ever/. Adrian Kingsley-Hughes, "HP Stream 7: Windows 8.1 Tablet for $79," *ZDNet*, February 9, 2015, http://www.zdnet.com/article/hp-stream-7-windows-8-1-tablet-for-79/; Sam Wakoba, "Jumia Starts Discount Sales for X-Touch Phonepads for Everyone in Nigeria," *TechMoran*, March 13, 2014, http://techmoran.com/jumia-starts-discount-sales-for-x-touch-phonepads-for-everyone-in-nigeria/.

78. Richard Heeks, "Raspberry Pi: A Paradigm Shift for ICT4D?," *ICTs for Development Blog*, October 29, 2012, http://ict4dblog.wordpress.com/2012/10/29/raspberry-pi-a-paradigm-shift-for-ict4d/.

79. OAfrica, "South African Desktop Computer Runs on Android, Uses 20 Watts of Power," *OAfrica.com*, March 10, 2014, http://www.oafrica.com/business/south-african-desktop-computer-runs-on-android-uses-20-watts-of-power/.

80. Lucia Terrenghi, Laura Garcia-Barrio, and Lidia Oshlyansky, "Tablets Use in Emerging Markets: An Exploration," in *Proceedings of the 15th International Conference on Human–Computer Interaction with Mobile Devices and Services—MobileHCI '13* (New York: ACM Press, 2013), 594–599, doi:10.1145/2493190.2494438.

81. Jonathan Nalder, "What the Post-PC Era Means for Education," *Educational Technology Debate* (blog), April 15, 2011, http://edutechdebate.org/tablet-computers-in-education/what-the-post-pc-era-means-for-education/; staff writer, "Tablet Teachers: Schools in Africa Are Going Digital—with Encouraging Results," *The Economist*, December 8, 2012, http://www.economist.com/news/business/21567972-schools-africa-are-going-digitalwith-encouraging-results-tablet-teachers; Evgeny Kaganer, Gabriel A. Giordano, Sebastien Brion, and Marco Tortoriello, "Media Tablets for Mobile Learning," *Communications of the ACM* 56, no. 11 (2013): 68–75, doi:10.1145/2500494.

82. Charles Kane, Walter Bender, Jody Cornish, and Neal Donahue, *Learning to Change the World: The Social Impact of One Laptop Per Child* (New York: Palgrave Macmillan, 2012).

83. K. Philip, L. Irani, and P. Dourish, "Postcolonial Computing: A Tactical Survey," *Science, Technology & Human Values* 37, no. 1 (November 21, 2010): 3–29, doi:10.1177/0162243910389594; Kenneth L. Kraemer, Jason Dedrick, and Prakul Sharma, "One Laptop per Child: Vision vs. Reality," *Communications of the ACM* 52, no. 6 (June 1, 2009): 66–73, doi:10.1145/1516046.1516063.

84. Caitlin Bentley, "The OLPC Laptop: Educational Revolution or Devolution?," in *Proceedings of the World Conference on E-Learning in Corporate, Government, Healthcare, and Higher Education* (Chesapeake, VA: Association for the Advancement of Computing in Education, 2007), 647–652; Ayodeji A. Fajebe, Michael L. Best, and Thomas N. Smyth, "Is the One Laptop Per Child Enough? Viewpoints from Classroom Teachers in Rwanda," *Information Technologies & International Development* 9, no. 3 (2013): 29–40.

85. Mark Warschauer and Morgan Ames, "Can One Laptop per Child Save the World's Poor?," *Journal of International Affairs* 64, no. 1 (2010): 33–51.

86. Ethan Zuckerman, "Decentralizing the Mobile Phone: A Second ICT4D Revolution?," *Information Technologies & International Development* 6 (2010): 99–103.

87. Sam Parales, "How Does the New Haitian Tablet Measure Up?," *ICTWorks* (blog), April 18, 2014, http://www.ictworks.org/2014/04/18/how-does-the-new-haitian-surtab-tablet-measure-up/; Emeka Okafor, "Qelasy—Education Tablet Computer," *Timbuktu Chronicles* (blog), September 21, 2014, http://timbuktuchronicles.blogspot.com/2014/09/qelasy-education-tablet-computer.html.

88. Lisa Horner, *A Human Rights Approach to the Mobile Internet* (Mellville, South Africa: Association for Progressive Communications, 2011), 16, https://www.apc.org/en/pubs/issue/human-rights-approach-mobile-internet.

89. Donner and Walton, "Your Phone Has Internet—Why Are You at a Library PC?"; François Bar, Chris Coward, Lucas Koepke, Chris Rothschild, Araba Sey, and George Sciadas, "The Impact of Public Access to ICTs," in *Proceedings of the Sixth International Conference on Information and Communication Technologies and Development Full Papers—ICTD '13*, vol. 1 (New York: ACM Press, 2013), 34–42, doi:10.1145/2516604.2516619.

90. Rohit Prasad and Rupamanjari Sinha Ray, "Telecentres Go Where Mobile Phones Fear to Tread: Evidence from India," in *Regional International Telecommunications Society India Conference* (New Delhi, 2012), 1–26; Wallace Chigona, O. Lekwane, Kim Westcott, and Agnes Chigona, "Uses, Benefits and Challenges of Public Access Points in the Face of Growth of Mobile Technology," *The Electronic Journal of Information Systems in Developing Countries* 49, no. 2006 (2011): 1–14; Donner and Walton, "Your Phone Has Internet—Why Are You at a Library PC?"; Isabella Rega, Sara Vannini, Miguel Raimilla, and Laia Fauró, "Telecentres and Mobile Technologies: A Global Pilot Study," *Associazione Seed Technical Report* (Canobbio, Switzerland: Seed, October 2013), http://seedlearn.org/telecentres-and-mobile-technologies-global-pilot-study.

91. Laura Forlano, "Anytime? Anywhere?: Reframing Debates Around Municipal Wireless Networking," *The Journal of Community Informatics* 4, no. 1 (2008), np, http://ci-journal.net/index.php/ciej/article/view/438/401.

92. The Broadband Commission, *The State of Broadband 2014: Broadband for All* (Geneva, Switzerland: ITU and UNESCO, 2014), http://www.broadbandcommission .org/Documents/reports/bb-annualreport2014.pdf.

93. Philip M. Napoli and Jonathan A. Obar, "The Emerging Mobile Internet Under-class: A Critique of Mobile Internet Access," *The Information Society* 30, no. 5 (October 2, 2014): 323–334, doi:10.1080/01972243.2014.944726; Horner, *A Human Rights Approach to the Mobile Internet*; Marion Walton, "Pavement Internet: Mobile Media Economies and Ecologies in South Africa," in *The Routledge Companion to Mobile Media*, ed. Gerard Goggin and Larissa Hjorth (London: Routledge, 2014), 450–461.

94. Naomi S. Baron, "Does Mobile Matter? The Case of One-off Reading," in *Routledge Companion to Mobile Media*, ed. Gerard Goggin and Larissa Hjorth (New York: Routledge, 2014), 225–235.

95. Timothy B. Lee, "The PC Is Dead, and This Year's CES Proves It," *The Washington Post*, January 8, 2014, http://www.washingtonpost.com/blogs/the-switch/wp/2014/ 01/08/the-pc-is-dead-and-this-years-ces-proves-it/; Raj Sabhok, "Death of the PC: Time to Kiss Your Computer Goodbye?," *Forbes.com*, August 13, 2013, http:// www.forbes.com/sites/rajsabhlok/2013/08/13/death-of-the-pc-time-to-kiss-your -computer-goodbye/.

96. Ina Fried, "Steve Jobs at D8: Post-PC Era Is Nigh," *CNET*, June 1, 2010, http:// news.cnet.com/8301-13860_3-20006526-56.html; Thanh Pham, "How I Ditched My Laptop for an iPad with a Few Apps and Accessories," *Lifehacker*, April 7, 2014, http://lifehacker.com/how-i-ditched-my-laptop-for-an-ipad-with-a-few-apps-and -1560205540.

Chapter 11

1. Jenna Burrell, "Is the Digital Divide a Defunct Framework?," *Global Policy* (blog), September 5, 2012, http://www.globalpolicyjournal.com/blog/05/09/2012/digital -divide-defunct-framework.

2. Harmeet Sawhney, "Innovations at the Edge: The Impact of Mobile Technologies on the Character of the Internet," in *Mobile Technologies: From Telecommunications to Media*, ed. Gerard Goggin and Larissa Hjorth (New York: Routledge, 2009), 105–117.

3. Tim Berners-Lee, Robert Cailliau, Ari Luotonen, Henrik Frystyk Nielsen, and Arthur Secret, "The World-Wide Web," *Communications of the ACM* 37, no. 8 (1994): 76–82, doi:10.1145/179606.179671.

4. Luigi Atzori, Antonio Iera, and Giacomo Morabito, "The Internet of Things: A Survey," *Computer Networks* 54, no. 15 (October 2010): 2787–2805, doi:10.1016/ j.comnet.2010.05.010.

5. Chris Anderson and Michael Wolff, "The Web Is Dead. Long Live the Internet," *Wired*, September 2010, http://www.wired.com/magazine/2010/08/ff_webrip/.

6. danah m. boyd and Nicole B. Ellison, "Social Network Sites: Definition, History, and Scholarship," *Journal of Computer-Mediated Communication* 13, no. 1 (October 17, 2007): 210–230, doi:10.1111/j.1083-6101.2007.00393.x.

7. Danny Crichton, "As Mobile Roars Ahead, It's Time to Finally Admit the Web Is Dying," *Techcrunch*, May 9, 2014, http://techcrunch.com/2014/05/09/as-mobile-roars-ahead-its-time-to-finally-admit-the-world-wide-web-is-dying/.

8. Sawhney, "Innovations at the Edge."

9. Cory Doctorow, *Information Doesn't Want to Be Free* (San Francisco: McSweeney's, 2014), 1.

10. Jonathan Lillie, "Cultural Access, Participation, and Citizenship in the Emerging Consumer-Network Society," *Convergence: The International Journal of Research into New Media Technologies* 11, no. 2 (June 1, 2005): 41–48, doi:10.1177/135485650501100205, p. 43.

11. Balancing Act Africa, "New Report Gives Insights on Mobile Apps Potentials in Africa," *Balancing Act Africa*, July 15, 2011, http://www.balancingact-africa.com/news/en/issue-no-563-0/top-story/new-report-gives-ins/en.

12. Vital Wave Consulting, "App-Happy New Year," *The Nugget* (newsletter), January 23, 2013, http://www.vitalwaveconsulting.com/newsletter/2013/1_23_13.htm.

13. Tarleton Gillespie, "The Politics of 'Platforms,'" *New Media & Society* 12, no. 3 (February 9, 2010): 347–364, doi:10.1177/1461444809342738.

14. John Battelle, "Put Your Taproot into the Independent Web," *Searchblog*, January 24, 2012, http://battellemedia.com/archives/2012/01/put-your-taproot-into-the-independent-web.php.

15. Matthew L. Smith and Laurent Elder, "Open ICT Ecosystems Transforming the Developing World," *Information Technologies and International Development* 6, no. 1 (2010): 65–71.

16. Ryan Gallagher, "India's Spies Want Data on Every BlackBerry Customer Worldwide," *Slate.com*, 2013, http://www.slate.com/blogs/future_tense/2013/02/22/india_wants_data_on_every_blackberry_customer_worldwide.html.

17. Irina Shklovski, Scott D. Mainwaring, Halla Hrund Skúladóttir, and Höskuldur Borgthorsson, "Leakiness and Creepiness in App Space: Perceptions of Privacy and Mobile App Use," in *Proceedings of the 32nd Annual ACM Conference on Human Factors in Computing Systems—CHI '14* (New York: ACM Press, 2014), doi:10.1145/2556288.2557421; Tony Bradley, "Location Tracking in Mobile Apps Is Putting Users at

Risk," *CSO Online*, January 16, 2015, http://www.csoonline.com/article/2871933/mobile-security/location-tracking-in-mobile-apps-is-putting-users-at-risk.html.

18. Amar Toor, "Why a Messaging App Meant for Festivals Became Massively Popular during Hong Kong Protests," *The Verge*, October 16, 2014, http://www.theverge.com/2014/10/16/6981127/firechat-messaging-app-accidental-protest-app-hong-kong; Jason Li, "Fact Checking the Hype around Mesh Networks and Fire-Chat," *88-Bar.com* (blog), January 11, 2015, http://www.88-bar.com/2015/01/fact-checking-the-hype-around-mesh-networks-and-firechat/.

19. Natasha Culzac, "Egypt's Police 'Using Social Media and Apps Like Grindr to Trap Gay People,'" *The Independant*, 2015, http://www.independent.co.uk/news/world/africa/egypts-police-using-social-media-and-apps-like-grindr-to-trap-gay-people-9738515.html.

20. Ronald Deibert, "Trouble at the Border: China's Internet," *Index on Censorship* 42, no. 2 (2013): 132–135, doi:10.1177/0306422013495334.

21. Abby Liu, "China's Crackdown on WeChat," *Global Voices*, March 14, 2014, http://globalvoicesonline.org/2014/03/14/chinas-crackdown-on-wechat/.

22. Katie Shilton, "Four Billion Little Brothers?: Privacy, Mobile Phones, and Ubiquitous Data Collection," *Commununications of the ACM* 7 (2009): 40–47, doi:10.1145/1592761.1592778.

23. Jonathan Zittrain, *The Future of the Internet and How to Stop It* (New Haven, CT: Yale University Press, 2008), 71–72.

24. Ibid., 3.

25. Ramon Lobato and Julian Thomas, "Informal Mobile Economies," in *Routledge Companion to Mobile Media*, ed. Gerard Goggin and Larissa Hjorth (New York: Routledge, 2014), 114–122.

26. Gary Sims, "How Cyanogen Plans to Be Android's Open-Source Champion," *Android Authority* (blog), January 16, 2014, http://www.androidauthority.com/cyanogen-androids-open-source-champion-336012/; Lobato and Thomas, "Informal Mobile Economies."

27. Alison Powell, "Openness and Enclosure in Mobile Internet Architecture," in *Theories of the Mobile Internet: Materialities and Imaginaries*, ed. Andrew Herman, Jan Hadlaw, and Thom Swiss (New York: Routledge, 2015), 25–44.

28. Marshini Chetty, Richard Banks, A. J. Bernheim Brush, Jonathan Donner, and Rebecca E. Grinter, "'You're Capped!' Understanding the Effects of Bandwidth Caps on Broadband Use in the Home," in *Proceedings of CHI 2012, May 5–10, 2010, Austin, TX* (New York: ACM, 2012), 3021–3030, doi:10.1145/2207676.2208714; Marshini Chetty, Srikanth Sundaresan, Sachit Muckaden, Nick Feamster, and Enrico Calandro, "Measuring Broadband Performance in South Africa," in *Proceedings of the 4th Annual*

Symposium on Computing for Development—ACM DEV-4 '13 (New York: ACM Press, 2013), 1–10, doi:10.1145/2537052.2537053.

29. Andrew Odlyzko, Bill St. Arnaud, Erik Stallman, and Michael Weinberg, "Know Your Limits: Considering the Role of Data Caps and Usage Based Billing in Internet Access Service," *Public Knowledge*, April 23, 2012, http://www.publicknowledge.org/documents/know-your-limits-considering-the-role-of-data-caps-and-usage-based-billing.

30. Nikolai Tillmann, Michal Moskal, Jonathan de Halleux, and Manuel Fahndrich, "TouchDevelop: Programming Cloud-Connected Mobile Devices via Touchscreen," in *Proceedings of the 10th SIGPLAN Symposium on New Ideas, New Paradigms, and Reflections on Programming and Software—ONWARD '11* (New York: ACM Press, 2011), 49–60, doi:10.1145/2048237.2048245.

31. Ethan Zuckerman, "Decentralizing the Mobile Phone: A Second ICT4D Revolution?," *Information Technologies & International Development* 6 (2010): 99–103; Ben Goldsmith, "The Smartphone App Economy and App Ecosystems," in *Routledge Companion to Mobile Media*, ed. Gerard Goggin and Larissa Hjorth (New York: Routledge, 2014), 171–180; Anderson and Wolff, "The Web Is Dead. Long Live the Internet."

32. Bryan Pon, Timo Seppälä, and Martin Kenney, "Android and the Demise of Operating System-based Power: Firm Strategy and Platform Control in the Post-PC World," *Telecommunications Policy* 38, no. 11 (2014): 979–991, doi:10.1016/j.telpol.2014.05.001; Bryan Pon, Timo Seppälä, and Martin Kenney, "One Ring to Unite Them All: Convergence, the Smartphone, and the Cloud," *Journal of Industry, Competition, and Trade* 15, no. 1 (2015), 21–33.

33. Gerard Goggin, "Google Phone Rising: The Android and the Politics of Open Source," *Continuum* 26, no. 5 (2012): 741–752, doi:10.1080/10304312.2012.706462; Sims, "How Cyanogen Plans to Be Android's Open-Source Champion."

34. Pon, Seppälä, and Kenney, "Android and the Demise of Operating System-based Power," 988.

35. Ibid.; Pon, Seppälä, and Kenney, "One Ring to Unite Them All."

36. Ben Eaton, Carsten Elaluf-Calderwood, Silvia Sorensen, and Youngjin Yoo, "Dynamic Structures of Control and Generativity in Digital Ecosystem Service Innovation: The Cases of the Apple and Google Mobile App Stores," *Working Paper Series, 183* (London: London School of Economics and Political Science, 2011), http://www.lse.ac.uk/management/documents/isig-wp/ISIG-WP-183.PDF.

37. Mark Surman, Corina Gardner, and David Ascher, "Local Content, Smartphones, and Digital Inclusion," *Innovations: Technology, Governance, Globalization* 9, no. 3–4 (July 2014): 63–74, doi:10.1162/inov_a_00217, p. 68.

38. Ibid.

39. Surman, Gardner, and Ascher, "Local Content, Smartphones, and Digital Inclusion."

40. Ibid.

41. Ibid., 77.

42. Harsha Liyanage, "Google's Android, Dis-Empowering the Poor?," *eNovation4D* (blog), March 29, 2013, http://enovation4d.blogspot.co.uk/2013/03/googles-android-dis-empowering-poor.html.

43. Jon Russell, "Google Reverses Plan to Stop Paying Android Developers in Argentina, Will Support Them 'for the Time Being,'" *The Next Web*, June 27, 2013, http://thenextweb.com/google/2013/06/27/google-reverses-plan-to-stop-paying-android-developers-in-argentina-will-support-them-for-the-time-being/.

44. L. J. Benn, "Mobile Data Services in the South African Wireless Industry: A New Value Chain," *International Journal of Technology, Policy* 5, no. 2 (2005): 121–131; Mark De Reuver, Tim De Koning, Harry Bouwman, and Wolter Lemstra, "How New Billing Processes Reshape the Mobile Industry," *Info* 11, no. 1 (2009): 78–93, doi:10.1108/14636690910933019.

45. Lillie, "Cultural Access, Participation, and Citizenship in the Emerging Consumer-Network Society."

46. Surman, Gardner, and Ascher, "Local Content, Smartphones, and Digital Inclusion," 71.

47. Deibert, "Trouble at the Border."

48. Sarah Wagner and Mireia Fernández-Ardèvol, "Local Content Production and the Political Economy of the Mobile App Industries in Argentina and Bolivia," *New Media & Society* (2015), published online before print, doi:10.1177/1461444815571112, p. 14.

49. Ibid.

50. Locke, "The Challenge of Sustaining App Entrepreneurs"; Goldsmith, "The Smartphone App Economy and App Ecosystems."

51. Russell Southwood, "The Irresistible Law of Circles: Getting Mobile and Online Content to African Users in a Fragmented Market," *Balancing Act Africa*, August 15, 2013, http://www.balancingact-africa.com/news/en/issue-no-668-0/top-story/the-irresistible-law/en.

52. David Souter, "Mobile Internet Usage and Demand in Kenya: The Experience of Early Adopters," *Making Broadband Accessible for All* (London: Vodafone Group, May 2011), http://www.vodafone.com/content/dam/vodafone/about/public_policy/

policy_papers/public_policy_series_12.pdf; Erik Hersman, "The Potential of Mobile Web Content in East Africa," in *Making Broadband Accessible for All* (London: Vodafone Group, 2011), 21–30.

53. Maja Andjelkovic and Saori Imaizumi, "Mobile Entrepreneurship and Employment," in *Information and Communications for Development 2012: Maximizing Mobile*, ed. World Bank, vol. 2011 (Washington, DC: World Bank, 2012), 75–86.

54. Banks, "An Inconvenient Truth?"; Christopher Foster and Richard Heeks, "Analyzing Policy for Inclusive Innovation: The Mobile Sector and Base-of-the-Pyramid Markets in Kenya," *Innovation and Development* 3, no. 1 (April 2013): 103–119, doi: 10.1080/2157930X.2013.764628; Christopher Foster and Richard Heeks, "Conceptualising Inclusive Innovation: Modifying Systems of Innovation Frameworks to Understand Diffusion of New Technology to Low-Income Consumers," *European Journal of Development Research*, April 4, 2013, 1–23, doi:10.1057/ejdr.2013.7; Erik Hersman, "Mobilizing Tech Entrepreneurs in Africa," *Innovations: Technology, Governance, Globalization* 7, no. 4 (2012): 59–68; Chris Locke, "The Challenge of Sustaining App Entrepreneurs," *Innovations: Technology, Governance, Globalization* 7, no. 4 (2012): 21–26, doi:10.1162/INOV_a_00149.

55. Richard Florida, "Toward the Learning Region," *Futures* 27, no. 5 (1995): 527–536; Michael E. Porter, *The Competitive Advantage of Nations* (New York: Free Press, 1990); Annalee Saxenian, *Regional Advantage: Culture and Competition in Silicon Valley and Route 128* (Cambridge, MA: Harvard University Press, 1994).

56. Danny Crichton, "Approaching the Frontier: How One Entrepreneur Is Building the Future in Myanmar," *Techcrunch*, February 24, 2014, http://techcrunch.com/2014/02/24/approaching-the-frontier-how-one-entrepreneur-is-building-the-future-in-myanmar/.

57. Banks, McDonald, and Scialom, "Mobile Technology and the Last Mile: 'Reluctant Innovation' and FrontlineSMS"; Pádraig Carmody, *New Socio-Economy in Africa? Thintegration and the Mobile Phone Revolution, The Institute for International Integration Studies Discussion Paper Series*, vol. 279 (Institute for International Integration Studies, 2009); Johan Hellström, *The Innovative Use of Mobile Applications in East Africa* (Stockholm, 2010); Nathan Amanquah and Mjumo Mzyece, "Mobile Application Research and Development: The African Context," in *Proceedings of the 2nd ACM Symposium on Computing for Development—ACM DEV '12* (New York: ACM Press, 2012), np, doi:10.1145/2160601.2160627.

58. Amol Sharma, "India, A New Facebook Testing Ground," *The Wall Street Journal*, October 20, 2012, http://online.wsj.com/news/articles/SB10000872396390443749204578048384116646940.

59. Emeka Okafor, "Able Wireless | Video Streaming with Raspberry Pi," *Timbuktu Chronicles* (blog), August 9, 2013, http://timbuktuchronicles.blogspot.com/2013/08/able-wireless-video-streaming-with.html.

60. Hersman, "Mobilizing Tech Entrepreneurs in Africa."

61. InfoDev, *The Business Models of mLabs and mHubs: An Evaluation of infoDev's Mobile Innovation Support Pilots* (Washington, DC, n.d.).

62. Jon Gosier, "What Tech Hubs Are Getting Wrong in Africa (and How to Fix It)," *AppAfrica*, 2014, http://blog.appfrica.com/2014/08/13/what-tech-hubs-are-getting -wrong-in-africa-and-how-to-fix-it/.

63. Loren Treisman, "Do Africa's Technology Entrepreneurs Need Charity?," *BBC News*, August 9, 2012, http://www.bbc.com/news/business-19195665.

64. Isse Lapowsky, "Zuckerberg Expands Internet.org After Net Neutrality Uproar," *Wired*, May 4, 2015, http://www.wired.com/2015/05/internet-org-expands-net -neutrality/.

65. Digital Fuel Monitor, "Google, Telcos and the Push for a Vertically Integrated Non-Neutral Internet—Friends, Not Foes," *Digital Fuel Monitor* (blog), November 2014, http://dfmonitor.eu/insights/2014_nov_premium_google/.

66. Internet desk, "Airtel Defends Airtel Zero, Calls It Toll-Free Service," *The Hindu*, April 20, 2015, http://www.thehindu.com/sci-tech/technology/airtels-stand-on-net-neutrality-and-airtel-zero/article7122394.ece.

67. Telecom Regulatory Authority of India, *Consultation Paper on Regulatory Framework for Over-the-Top (OTT) Services No. 2/2015*, New Delhi, 2015, http://www.trai .gov.in/WriteReaddata/ConsultationPaper/Document/OTT-CP-27032015.pdf.

68. Anita Babu and Surabhi Agarwal, "What the One Million Voices Said to Trai," *Business Standard*, New Delhi, April 29, 2015, http://www.business-standard.com/ article/current-affairs/what-the-one-million-voices-said-to-trai-115042900984_1 .html.

69. Mark Zuckerberg, Untitled Facebook Post, May 4, 2015, https://www .facebook.com/zuck/videos/10102066901270081/?pnref=story.

70. Internet.org, "Announcing the Internet.org Platform," Internet.org Press Releases, May 4, 2015, https://www.internet.org/press/announcing-the-internet-dot -org-platform.

71. Tarleton Gillespie, "The Politics of 'Platforms,'" *New Media & Society* 12, no. 3 (February 9, 2010): 347–364, doi:10.1177/1461444809342738; Bryan Pon, "Facebook 'Levelling' the Playing Field?," *Caribou Digital Blog*, May 11, 2015, http:// cariboudigital.net/facebook-levelling-the-playing-field/.

72. LIRNEasia, "Comments by LIRNEasia on the Consultation Paper on Regulatory Framework for Over-the-Top (OTT) Services, Submitted to the Telecom Regulatory Agency of India on 24th April 2015," Colombo, Sri Lanka, 2015, http://lirneasia.net/ wp-content/uploads/2015/04/LIRNEasia_Response_TRAIOTT_V2.pdf.

73. Craig Wilson, "Crackdown Irks MTN BlackBerry Clients," *Techcentral.co.za*, November 22, 2012, http://www.techcentral.co.za/crackdown-irks-mtn-blackberry -clients/.

74. Jeremy Malcolm, "Net Neutrality and the Global Digital Divide," *Electronic Frontier Foundation* (blog), July 24, 2014, https://www.eff.org/deeplinks/2014/07/ net-neutrality-and-global-digital-divide.

75. Christopher T. Marsden, "Network Neutrality: A Research Guide," in *Research Handbook on Governance of the Internet.*, ed. Ian Brown (Cheltenham, UK: Edward Elgar, 2013), 419–444, doi:10.4337/9781849805049.00026.

76. Barbara vann Schewick, *Internet Architecture and Innovation* (Cambridge, MA: MIT Press, 2010).

77. Susan Crawford, "Zero for Conduct," *Medium*, January 7, 2015, https://medium .com/backchannel/less-than-zero-199bcb05a868.

78. Fred Wilson, "The Scourge Of Zero Rating," *AVC* (blog), July 31, 2014, http:// avc.com/2014/07/the-scourge-of-zero-rating/.

79. Steve Song, "Globalising the Net Neutrality Debate," *Medium*, December 11, 2014,   https://medium.com/@stevesong/globalising-the-net-neutrality-debate -43d339f575b3.

80. Helani Galpaya, cited in "WS208—Net Neutrality, Zero Rating and Develop-ment: What's the Data?," Rough Transcript, *Internet Governance Forum 2014* (Istan-bul, Turkey, September 3, 2014), http://www.intgovforum.org/cms/174-igf-2014/ transcripts/1969-2014-09-03-ws208-net-neutrality-zero-rating-and-development -room-5.

81. Kul Wadhwa and Howie Fung, "Converting Western Internet to Indigenous Internet: Lessons from Wikipedia," *Innovations: Technology, Governance, Globalization* 9, no. 3–4 (July 2014): 132–141, doi:10.1162/inov_a_00224, p. 138.

82. Gregory Wang, "Opera News Facebook Joins Opera in Initiative to Bring Inter-net to the Next Billion Users," *Opera News*, June 5, 2014, http://blogs.opera.com/ news/2014/06/facebook-joins-opera-initiative-bring-internet-next-billion-users/.

83. Ingrid Lunden, "FreedomPop to Offer App-Sized Data Plans, Free Use of Spon-sored Apps," *Techcrunch*, June 24, 2014, http://techcrunch.com/2014/06/24/ freedompop-to-offer-app-sized-data-plans-free-use-of-sponsored-apps/?ncid=rss.

84. Jack Clarke, "'Press Like for 1MB of Data': Facebook Picks up Pryte in-App Data Tech," *The Register*, June 3, 2012, http://www.theregister.co.uk/2014/06/03/ facebook_pryte_acquisition/.

85. Ryan Knutson, "Sprint Will Sell a $12 Wireless Plan That Only Connects to Facebook or Twitter," *The Wall Street Journal*, July 30, 2014, http://blogs.wsj.com/ digits/2014/07/30/sprint-tries-a-facebook-only-wireless-plan/.

86. Internet.org, "Participation Guidelines," Internet.org Developer Resources, accessed May 11, 2015, https://developers.facebook.com/docs/internet-org/participation-guidelines.

87. Michael L. Best, "The Internet That Facebook Built," *Communications of the ACM* 57, no. 12 (2014): 21–23, doi:10.1145/2676857.

88. Nancy Scola, "Will Apps That Don't Burn through Your Data Plan Destroy the Internet or Save It?," *The Washington Post*, August 15, 2014, http://www.washingtonpost.com/blogs/the-switch/wp/2014/08/15/will-apps-that-dont-burn-through-your-data-plan-destroy-the-internet-or-save-it/.

89. Eli Dourado, "How Net Neutrality Hurts the Poor," *The Ümlaut*, April 30, 2014, https://theumlaut.com/2014/04/30/how-net-neutrality-hurts-the-poor/; Helani Galpaya, cited in "WS208—Net Neutrality, Zero Rating and Development: What's the Data?," Rough Transcript.

90. Ian Scales, "Slovenia Bans 'Zero-Rating' Price Discrimination as the EU Dithers," *Telecom TV*, January 25, 2015, http://www.telecomtv.com/articles/net-neutrality/slovenia-bans-zero-rating-price-discrimination-as-the-eu-dithers-12125/; David Meyer, "In Chile, Mobile Carriers Can No Longer Offer Free Twitter, Facebook or WhatsApp," *GigaOm*, May 28, 2014, https://gigaom.com/2014/05/28/in-chile-mobile-carriers-can-no-longer-offer-free-twitter-facebook-and-whatsapp/.

91. Steve Song, "A Better Approach to Zero-Rating," *Many Possibilities*, November 13, 2014, https://manypossibilities.net/2014/11/a-better-approach-to-zero-rating/.

92. Jana.com, "Product," *Jana Website*, accessed May 11, 2015, https://jana.com/product.

93. Linnet Taylor, "The Big Squeeze: Why the Mobile Internet Isn't Enough," *Doubt Wisely, and Never Lose Your Penguin*, October 29, 2013, http://linnettaylor.wordpress.com/2013/10/29/the-big-squeeze-why-the-mobile-internet-isnt-enough/.

94. Rohan Samarajiva, "A Response to an Essay about Mobile Access Not Being Real Access to the Internet," *LIRNEasia*, November 5, 2013, http://lirneasia.net/2013/11/a-response-to-an-essay-about-mobile-access-not-being-real-access-to-the-internet/.

95. ITforChange and IDRC, "Roundtable on Inclusion in the Network Society: Mapping Development Alternatives, Forging Research Agendas," *Workshop Report*, September 29, 2014, 28, http://itforchange.net/sites/default/files/RoundTableonInclusionintheNetworkSociety-FullReport.pdf.

Chapter 12

1. Mark Warschauer, *Technology and Social Inclusion: Rethinking the Digital Divide* (Cambridge, MA: MIT Press, 2003).

2. Daniel Miller and Don Slater, *The Internet: An Ethnographic Approach* (Oxford, UK: Berg, 2000).

3. Assa Doron and Robin Jeffrey, *The Great Indian Phone Book: How the Cheap Cell Phone Changes Business, Politics, and Daily Life* (Cambridge, MA: Harvard University Press, 2013).

4. Amanda Watson, "Mobile Phones and Media Use in Madang Province of Papua New Guinea," *Pacific Journalism Review* 19, no. 2 (2013): 156–175.

5. Neil Gough, "Chinese Now Prefer Mobile When Going Online," *The New York Times Online*, July 22, 2014, http://sinosphere.blogs.nytimes.com/2014/07/22/smartphones-surpass-computers-for-internet-use-in-china/. See also CINIC, *Statistical Report on Internet Development in China* (Beijing, 2014), http://www1.cnnic.cn/IDR/ReportDownloads/201411/P020141102574314897888.pdf.

6. Rodney Wai-chi Chu, Leopoldina Fortunati, Pui-Lam Law, and Shanhua Yang, eds., *Mobile Communication and Greater China* (London: Routledge, 2012).

7. Jack Linchuan Qiu, "Network Societies and Internet Studies: Rethinking Time, Space, and Class," in *The Oxford Handbook of Internet Studies*, ed. William H. Dutton (Oxford: Oxford University Press, 2013), 110–128; Jack Linchuan Qiu, "Working-Class ICTs, Migrants, and Empowerment in South China," *Asian Journal of Communication* 18, no. 4 (December 2008): 333–347, doi:10.1080/01292980802344232.

8. Cara J. Wallis, "Mobile Phones without Guarantees: The Promises of Technology and the Contingencies of Culture," *New Media & Society* 13, no. 3 (March 25, 2011): 471–485, doi:10.1177/1461444810393904; Cara J. Wallis, "New Media Practices in China: Youth Patterns, Processes, and Politics," *International Journal of Communication* 5 (2011): 406–436; Cara J. Wallis, *Technomobility in China: Young Migrant Women and Mobile Phones* (New York: New York University Press, 2013).

9. Elisa Oreglia, "When Technology Doesn't Fit," in *Proceedings of the Sixth International Conference on Information and Communication Technologies and Development Full Papers—ICTD '13*, vol. 1 (New York: ACM, 2013), 165–176, doi:10.1145/2516604.2516610.

10. Pui-lam Law, ed., *New Connectivities in China: Virtual, Actual and Local Interactions* (Berlin; Heidelberg: Springer, 2012).

11. Fie Velghe, "Lessons in Textspeak from Sexy Chick: Supervernacular Literacy in South African Instant and Text Messaging," in *African Literacies: Scripts, Ideologies, Education*, ed. Yonas Mesfun Asfaha and Ashraf K. Abdelhay (Newcastle-upon-Tyne: Cambridge Scholars, 2011), 63–87.

12. James E. Katz and Mark Aakhus, "Conclusion: Making Meaning of Mobiles: A Theory of Apparatgeist," in *Perpetual Contact: Mobile Communication, Private Talk,*

and Public Performance, ed. James E. Katz and Mark Aakhus (Cambridge, UK: Cambridge University Press, 2002), 301–318.

13. Sirpa Tenhunen, "Mobile Technology in the Village: ICTs, Culture, and Social Logistics in India," *Journal of the Royal Anthropological Institute* 14, no. 3 (September 2008): 515–534, doi:10.1111/j.1467-9655.2008.00515.x; M. Pathak-Shelat and C. DeShano, "Digital Youth Cultures in Small Town and Rural Gujarat, India," *New Media & Society* 16, no. 6 (2013): 983–1001, doi: 10.1177/1461444813496611.

14. Pádraig Carmody, *New Socio-Economy in Africa? Thintegration and the Mobile Phone Revolution, The Institute for International Integration Studies Discussion Paper Series*, vol. 279 (Trinity College, Dublin: Institute for International Integration Studies, 2009); Albrecht Fehske, Gerhard Fettweis, Jens Malmodin, and Gergely Biczok, "The Global Footprint of Mobile Communications: The Ecological and Economic Perspective," *IEEE Communications Magazine* 49, no. 8 (August 2011): 55–62, doi:10.1109/MCOM.2011.5978416.

15. Luigi Atzori, Antonio Iera, and Giacomo Morabito, "The Internet of Things: A Survey," *Computer Networks* 54, no. 15 (October 2010): 2787–2805, doi:10.1016/j.comnet.2010.05.010.

16. Martin Hilbert, "Big Data for Development: From Information to Knowledge Societies," *Available at SSRN* (2013), doi:10.2139/ssrn.2205145. Vanessa Frias-Martinez Enrique Frias-Martinze, "Enhancing Public Policy Decision Making Using Large-Scale Cell Phone Data," *UN Global Pulse* (blog), September 4, 2012, http://www.unglobalpulse.org/publicpolicyandcellphonedata.

17. Manuel Castells, *The Rise of the Network Society* (Malden, MA: Blackwell Publishing, 1996).

18. Vasja Vehovar, Pavle Sicherl, Tobias Hüsing, and Vesna Dolnicar, "Methodological Challenges of Digital Divide Measurements," *The Information Society* 22, no. 5 (December 2006): 279–290, doi:10.1080/01972240600904076; Jan van Dijk and Kenneth L. Hacker, "The Digital Divide as a Complex and Dynamic Phenomenon," *The Information Society* 19, no. 4 (September 1, 2003): 315–326, doi:10.1080/01972240309487; Amanda Lenhart and John B. Horrigan, "Re-Visualizing the Digital Divide as a Digital Spectrum," *IT & Society* 1, no. 5 (2003): 23–39; Warschauer, *Technology and Social Inclusion*; Eszter Hargittai and Gina Walejko, "The Participation Divide: Content Creation and Sharing in the Digital Age," *Information, Communication & Society* 11, no. 2 (March 2008): 239–256, doi:10.1080/13691180801946150; Siobhan Stevenson, "Digital Divide: A Discursive Move Away from the Real Inequities," *The Information Society* 25, no. 1 (January 13, 2009): 1–22, doi:10.1080/01972240802587539.

19. Roger Harris, "Is the Digital Divide about to Be Closed?," Roger Harris Associates blog, October 16, 2012, http://www.rogharris.org/4/post/2012/10/is-the-digital-divide-about-to-be-closed.html.

20. David Nemer, Shad Gross, and Nic True, "Materializing Digital Inequalities: The Digital Artifacts of the Marginalized in Brazil," in *Proceedings of the Sixth International Conference on Information and Communications Technologies and Development Notes—ICTD '13*, vol. 2 (New York: ACM Press, 2013), 108–111, doi:10.1145/2517899 .2517915.

21. Gary W. Pritchard and John Vines, "Digital Apartheid: An Ethnographic Account of Racialised HCI in Cape Town Hip-Hop," in *Proceedings of the SIGCHI Conference on Human Factors in Computing Systems—CHI '13* (New York: ACM Press, 2013), 2537–2546, doi:10.1145/2470654.2481350; Craig Wilson, "Digital Apartheid: Separate and Unequal," *Techcentral.co.za*, June 25, 2013, http://www.techcentral .co.za/digital-apartheid-separate-and-unequal/41373/.

22. Jamilah King, "How Big Telecom Used Smartphones to Create a New Digital Divide," *Colorlines.com*, December 6, 2011, http://colorlines.com/archives/2011/12/ the_new_digital_divide_two_separate_but_unequal_internets.html.

23. Jerry Watkins, Kathi R. Kitner, and Dina Mehta, "Mobile and Smartphone Use in Urban and Rural India," *Continuum* 26, no. 5 (October 2012): 685–697, doi:10.1080/10304312.2012.706458.

24. Marion Walton, "Mobile Literacies: Messaging, Txt and Social Media in m4Lit," in *Multimodal Approaches to Research and Pedagogy: Recognition, Resources and Access*, ed. A. Archer and D. Newfield (London: Routledge, 2014), 108–127.

25. Nora Young, *The Virtual Self: How Our Digital Lives Are Altering the World Around Us* (Toronto: McClelland & Stewart, 2012).

26. E. Hagen, "Mapping Change: Community Information Empowerment in Kibera," *Innovations* 6, no. 1 (2011): 69–94, doi:10.1162/INOV_a_00059.

27. Jenny Davis, "We Don't Have Data, We Are Data," *Cyborgology*, December 22, 2014, http://thesocietypages.org/cyborgology/2014/12/22/we-dont-have-data-we -are-data/.

28. James E. Katz and Chih-Hui Lai, "Mobile Locative Media," in *Routledge Companion to Mobile Media*, ed. Gerard Goggin and Larissa Hjorth (New York: Routledge, 2014), 53–62.

29. Qiu, *Working-Class Network Society*.

30. Ina Fried, "Steve Jobs at D8: Post-PC Era Is Nigh," *CNET*, June 1, 2010, http:// news.cnet.com/8301-13860_3-20006526-56.html.

31. Fahim Kawsar and A. J. Bernheim Brush, "Home Computing Unplugged," in *Proceedings of the 2013 ACM International Joint Conference on Pervasive and Ubiquitous Computing—UbiComp '13* (New York: ACM, 2013), 627–636, doi:10.1145/2493432 .2493494.

32. Alexis C. Madrigal, "Bruce Sterling on Why It Stopped Making Sense to Talk About 'The Internet' in 2012," *The Atlantic*, December 27, 2012, http://www .theatlantic.com/technology/archive/2012/12/bruce-sterling-on-why-it-stopped -making-sense-to-talk-about-the-internet-in-2012/266674/.

33. Gerard Goggin, "Facebook's Mobile Career," *New Media & Society* 16, no. 7 (July 21, 2014): 1068–1086, doi:10.1177/1461444814543996.

34. "Over-the-Top Phone Services: Joyn Them or Join Them," *Economist*, August 11, 2012, http://www.economist.com/node/21560298?fsrc=scn/tw_ec/joyn_them_or_ join_them; Tina George Karippacheril, Fatemeh Nikayin, Mark de Reuver, and Harry Bouwman, "Serving the Poor: Multisided Mobile Service Platforms, Openness, Competition, Collaboration and the Struggle for Leadership," *Telecommunications Policy* 37, no. 1 (February 2013): 24–34, doi:10.1016/j.telpol.2012.06.001.

35. Matthew Guilford, "To the Next Billion: Mobile Network Operators and the Content Distribution Value Chain," *Innovations: Technology, Governance, Globalization* 9, no. 3–4 (July 2014): 21–31, doi:10.1162/inov_a_00213.

36. Richard Handford, "Vodafone CEO: Facebook Is Asking for Special Treatment; $30–40B M&A Stash," *Mobile World Live*, February 11, 2014, http://www .mobileworldlive.com/vodafone-ceo-facebook-asking-special-treatment-30-40b-ma -stash; Guilford, "To the Next Billion."

37. Jennifer Haroon, "Joining Forces to Advocate for a More Affordable Internet," *Google Africa Blog*, October 19, 2013, http://google-africa.blogspot.com/2013/10/ joining-forces-to-advocate-for-more.html.

38. Steve Song, "Corporate Narratives, ICTs, and Development," *Many Possibilities*, January 2013, http://manypossibilities.net/2013/01/corporate-narratives-icts-and -development/.

39. Facebook, "Internet.org Announces SocialEDU," *Facebook Newsroom*, February 24, 2014, http://newsroom.fb.com/news/2014/02/internet-org-announces -socialedu/.

40. Nicola Green, Richard Harper, G. Murtagh, and Geoff Cooper, "Configuring the Mobile User: Sociological and Industry Views," *Personal and Ubiquitous Computing* 5, no. 2 (July 1, 2001): 146–156, doi:10.1007/s007790170017; Hector Postigo, "Questioning the Web 2.0 Discourse: Social Roles, Production, Values, and the Case of the Human Rights Portal," *The Information Society* 27, no. 3 (May 2011): 181–193, doi :10.1080/01972243.2011.566759; Jonathan Grudin and John Pruitt, "Personas, Participatory Design and Product Development: An Infrastructure for Engagement," online *Proceedings of Participatory Design Conference 2002*, 144–161, http://ojs.ruc.dk/ index.php/pdc/article/view/249/241.

41. Erik Hersman, "Confusing ICT4D Practice with the Tech That Is Used," *Ushahidi Blog* (Nairobi, April 7, 2010), http://blog.ushahidi.com/index.php/2010/04/07/confusing-ict4d-practice-with-the-tech-that-is-used/; Anita Gurumurthy, "From Social Enterprises to Mobiles—Seeking a Peg to Hang a Premeditated ICTD Theory in an Ecology of Unequal Actors," *Information Technologies & International Development* 6, no. SE (2010): 57–63; Ethan Zuckerman, "Decentralizing the Mobile Phone: A Second ICT4D Revolution?," *Information Technologies & International Development* 6 (2010): 99–103; Steve Song, "Unpacking Our Mobile Broadband Future," *Many Possibilities*, October 15, 2012, http://manypossibilities.net/2012/10/unpacking-our-mobile-broadband-future/.

42. Qiu, *Working-Class Network Society*.

43. Astra Taylor, *The People's Platform: Taking Back Power and Culture in the Digital Age* (New York: Metropoolitan Books, 2014).

44. Lisa Horner, *A Human Rights Approach to the Mobile Internet* (Mellville, South Africa: Association for Progressive Communications, 2011), https://www.apc.org/en/pubs/issue/human-rights-approach-mobile-internet.

45. Francis Augusto Medeiros and Lee A. Bygrave, "Brazil's Marco Civil Da Internet: Does It Live up to the Hype?," *Computer Law & Security Review* 31, no. 1 (2015): 120–130, doi:10.1016/j.clsr.2014.12.001.

46. Peter Lunenfeld, *The Secret War between Uploading and Downloading* (Cambridge, MA: MIT Press, 2011).

47. Andrew Grantham and George Tsekouras, "Information Society: Wireless ICTs' Transformative Potential," *Futures* 36, no. 3 (April 2004): 359–377, doi:10.1016/S0016-3287(03)00066-1.

48. Mark Surman, Corina Gardner, and David Ascher, "Local Content, Smartphones, and Digital Inclusion," *Innovations: Technology, Governance, Globalization* 9, no. 3–4 (July 2014): 63–74, doi:10.1162/inov_a_00217.

49. Jonathan Donner, "Framing M4D: The Utility of Continuity and the Dual Heritage of 'Mobiles and Development,'" *The Electronic Journal of Information Systems in Developing Countries* 44, no. 3 (2010): 1–16; Jonathan Donner, Katrin Verclas, and Kentaro Toyama, "Reflections on MobileActive08 and the M4D Landscape," in *Proceedings of the First International Conference on M4D*, ed. John Sören Pettersson (Karlstad, Sweden: Karlstad University Studies, 2008), 73–83; Katy E. Pearce, "Phoning It In: Theory in Mobile Media and Communication in Developing Countries," *Mobile Media & Communication* 1, no. 1 (January 1, 2013): 76–82, doi:10.1177/2050157912459182; Richard A. Duncombe, "Mobiles for Development Research: Quality and Impact," in *Proceedings of the 2nd International Conference on M4D: Mobile Communication Technology for Development*, ed. Jakob Svensson and Gudrun Wicander (Karlstad, Sweden: Karlstad Universitet, 2010), 49–59.

50. Heather A Horst, "The Infrastructures of Mobile Media: Towards a Future Research Agenda," *Mobile Media & Communication* 1, no. 1 (2013): 147–152, doi:10.1177/2050157912464490; Jason Farman, "The Materiality of Locative Media on the Invisible Infrastructure of Mobile Networks," in *Theories of the Mobile Internet: Materialities and Imaginaries*, ed. Andrew Herman, Jan Hadlaw, and Thom Swiss (New York: Routledge, 2015): 45–59.

51. Ithiel de Sola Pool, "Introduction," *The Social Impact of the Telephone,* ed. Ithiel de Sola Pool (Cambridge, MA: MIT Press, 1977), 1–9.

52. Jerry Watkins, Kathi R. Kitner, and Dina Mehta, "Mobile and Smartphone Use in Urban and Rural India," *Continuum* 26, no. 5 (October 2012): 685–697, doi:10.1080/10304312.2012.706458; Gary Marsden, "What Is the Mobile Internet?," *Interactions* 14, no. 6 (November 2007): 24–25, doi:10.1145/1300655.1300672; Darja Groselj and Grant Blank, "Comparing Mobile and Non-Mobile Internet Users: How Are Mobile Users Different?," paper presented at the 2013 ICA Preconference on Mobile Communication, *10 Years On: Looking Forward in Mobile ICT Research*, May 17, 2013, London; Wallace Chigona, Darry Beukes, Junaid Vally, and Maureen Tanner, "Can Mobile Internet Help Alleviate Social Exclusion in Developing Countries?," *The Electronic Journal of Information Systems in Developing Countries* 36 (2009): 1–16.

53. Marion Walton, "Pavement Internet: Mobile Media Economies and Ecologies in South Africa," in *Routledge Companion to Mobile Media*, ed. Gerard Goggin and Larissa Hjorth (London: Routledge, 2014), 450–461; Jonathan Donner and Marion Walton, "Your Phone Has Internet—Why Are You at a Library PC? Re-Imagining Public Access for the Mobile Internet Era," in *Proceedings of INTERACT 2013: 14th IFIP TC 13 International Conference* (Berlin: Springer, 2013), 347–364, doi:10.1007/978-3-642-40483-2_25; Marion Walton and Pierrinne Leukes, "Prepaid Social Media and Mobile Discourse in South Africa," *Journal of African Media Studies* 5, no. 2 (2013): 149–167, doi:10.1386/jams.5.2.149_1; Marion Walton and Nicola Pallitt, "Grand Theft South Africa: Games, Literacy and Inequality in Consumer Childhoods," *Language and Education* 26, no. 4 (July 2012): 347–361, doi:10.1080/09500782.2012.691516; Marion Walton, "Social Distance, Mobility and Place: Global and Intimate Genres in Geo-Tagged Photographs of Guguletu, South Africa," in *Proceedings of the 8th ACM Conference on Designing Interactive Systems—DIS '10* (New York: ACM Press, 2010), 35–38, doi:10.1145/1858171.1858178; Marion Walton and Jonathan Donner, "Read-Write-Erase: Mobile-Mediated Publics in South Africa's 2009 Elections," in *Mobile Communication: Dimensions of Social Policy*, ed. James E. Katz (New Brunswick, NJ: Transaction Publishers, 2011), 117–132.

54. Katy E. Pearce, "Phoning It In: Theory in Mobile Media and Communication in Developing Countries," *Mobile Media & Communication* 1, no. 1 (January 1, 2013): 76–82, doi:10.1177/2050157912459182; Katy E. Pearce and Ronald E. Rice, "Digital Divides from Access to Activities: Comparing Mobile and Personal Computer

Internet Users," *Journal of Communication* 63, no. 4 (2013): 721–744, doi:10.1111/jcom.12045; Katy E. Pearce, Janine Slaker, and Nida Ahmad, "Is Your Web Everyone's Web? Theorizing the Web through the Lens of the Device Divide," *Theorizing the Web 2012*, April 12, 2012, http://fr.slideshare.net/katyp1/katy-pearce-ttw12.

55. Zuckerman, "Decentralizing the Mobile Phone."

56. Song, "Unpacking Our Mobile Broadband Future"; Song, "Corporate Narratives, ICTs, and Development."

57. Harris, "Is the Digital Divide about to Be Closed?"

58. Maeve Duggan and Aaron Smith, "Cell Internet Use 2013," *Pew Research Center Reports*, September 16, 2013, http://www.pewinternet.org/files/old-media//Files/Reports/2013/PIP_CellInternetUse2013.pdf.

59. Horner, *A Human Rights Approach to the Mobile Internet.*

60. Philip M. Napoli and Jonathan A. Obar, "The Emerging Mobile Internet Underclass: A Critique of Mobile Internet Access," *The Information Society* 30, no. 5 (October 2, 2014): 323–334, doi:10.1080/01972243.2014.944726.

61. Rohan Samarajiva, "Leveraging the Budget Telecom Network Business Model to Bring Broadband to the People," *Information Technologies & International Development* 6, no. SE (2010): 93–97; Ayesha Zainudeen, "Are the Poor Stuck in Voice? Conditions for Adoption of More-than-Voice Mobile Services," *Information Technologies & International Development* 7, no. 3 (2011): 45–59.

62. Emmanuel Forestier, Jeremy Grace, and Charles Kenny, "Can Information and Communication Technologies Be Pro-Poor?," *Telecommunications Policy* 26, no. 11 (December 2002): 623–646, doi:10.1016/S0308-5961(02)00061-7, p. 631.

63. Human Rights Council (HRC) of the United Nations, *The Promotion, Protection and Enjoyment of Human Rights on the Internet*, 2012, http://ap.ohchr.org/documents/E/HRC/d_res_dec/A_HRC_20_L13.doc.

64. Michael Gurstein, "Internet Justice: A Meme Whose Time Has Come," *Gurstein's Community Informatics Blog*, November 27, 2013, http://gurstein.wordpress.com/2013/11/27/internet-justice-a-meme-whose-time-has-come/.

Index